THE ROLE OF SMALL SATELLITES IN NASA AND NOAA EARTH OBSERVATION PROGRAMS

Committee on Earth Studies
Space Studies Board
Commission on Physical Sciences, Mathematics, and Applications
National Research Council

NATIONAL ACADEMY PRESS
Washington, D.C.

NOTICE: The project that is the subject of this report was approved by the Governing Board of the National Research Council, whose members are drawn from the councils of the National Academy of Sciences, the National Academy of Engineering, and the Institute of Medicine. The members of the committee responsible for the report were chosen for their special competences and with regard for appropriate balance. Support for this project was provided by Contract NASW 96013 between the National Academy of Sciences and the National Aeronautics and Space Administration, and was funded in part by a contract from the National Oceanic and Atmospheric Administration. Any opinions, findings, conclusions, or recommendations expressed in this material are those of the authors and do not necessarily reflect the views of the sponsor.

International Standard Book Number 0-309-06982-3

Copies of this report are available free of charge from:
Space Studies Board
National Research Council
2101 Constitution Avenue, NW
Washington, DC 20418
Copyright 2000 by the National Academy of Sciences. All rights reserved.

Printed in the United States of America

THE NATIONAL ACADEMIES

National Academy of Sciences
National Academy of Engineering
Institute of Medicine
National Research Council

The **National Academy of Sciences** is a private, nonprofit, self-perpetuating society of distinguished scholars engaged in scientific and engineering research, dedicated to the furtherance of science and technology and to their use for the general welfare. Upon the authority of the charter granted to it by the Congress in 1863, the Academy has a mandate that requires it to advise the federal government on scientific and technical matters. Dr. Bruce M. Alberts is president of the National Academy of Sciences.

The **National Academy of Engineering** was established in 1964, under the charter of the National Academy of Sciences, as a parallel organization of outstanding engineers. It is autonomous in its administration and in the selection of its members, sharing with the National Academy of Sciences the responsibility for advising the federal government. The National Academy of Engineering also sponsors engineering programs aimed at meeting national needs, encourages education and research, and recognizes the superior achievements of engineers. Dr. Wm. A. Wulf is president of the National Academy of Engineering.

The **Institute of Medicine** was established in 1970 by the National Academy of Sciences to secure the services of eminent members of appropriate professions in the examination of policy matters pertaining to the health of the public. The Institute acts under the responsibility given to the National Academy of Sciences by its congressional charter to be an adviser to the federal government and, upon its own initiative, to identify issues of medical care, research, and education. Dr. Kenneth I. Shine is president of the Institute of Medicine.

The **National Research Council** was organized by the National Academy of Sciences in 1916 to associate the broad community of science and technology with the Academy's purposes of furthering knowledge and advising the federal government. Functioning in accordance with general policies determined by the Academy, the Council has become the principal operating agency of both the National Academy of Sciences and the National Academy of Engineering in providing services to the government, the public, and the scientific and engineering communities. The Council is administered jointly by both Academies and the Institute of Medicine. Dr. Bruce M. Alberts and Dr. Wm. A. Wulf are chairman and vice chairman, respectively, of the National Research Council.

www.national-academies.org

COMMITTEE ON EARTH STUDIES

MARK R. ABBOTT, Oregon State University, *Chair*
OTIS B. BROWN,** University of Miami
JOHN R. CHRISTY, University of Alabama, Huntsville
CATHERINE GAUTIER, University of California, Santa Barbara
DANIEL J. JACOB,** Harvard University
CHRIS J. JOHANNSEN,* Purdue University
CHRISTOPHER O. JUSTICE, University of Virginia
VICTOR V. KLEMAS,* University of Delaware
BRUCE D. MARCUS,** TRW
M. PATRICK MCCORMICK,** Hampton University
ARAM M. MIKA,* Lockheed Martin Missiles and Space
RALPH F. MILLIFF, National Center for Atmospheric Research
RICHARD K. MOORE,* University of Kansas
SCOTT PACE, Rand
DALLAS L. PECK, U.S. Geological Survey (retired)
MICHAEL J. PRATHER, University of California, Irvine
R. KEITH RANEY, Johns Hopkins University Applied Physics Laboratory
DAVID T. SANDWELL, Scripps Institution of Oceanography
LAWRENCE C. SCHOLZ, West Orange, New Jersey
CARL F. SCHUELER, Raytheon Santa Barbara Remote Sensing
WALTER S. SCOTT,* EarthWatch
GRAEME L. STEPHENS, Colorado State University
KATHRYN D. SULLIVAN,* Columbus Ohio's Center of Science and Industry
FAWWAZ T. ULABY, University of Michigan
SUSAN L. USTIN, University of California, Davis
FRANK J. WENTZ, Remote Sensing Systems
THOMAS T. WILHEIT, JR.,* Texas A&M University
EDWARD F. ZALEWSKI, University of Arizona
ARTHUR A. CHARO, Senior Program Officer
INA B. ALTERMAN, Senior Program Officer
CARMELA J. CHAMBERLAIN, Senior Project Assistant (through March 1999)
THERESA M. FISHER, Senior Project Assistant (from April 1999)

---

* Term ended in 1998.
** Term ended in 1999.

SPACE STUDIES BOARD

CLAUDE R. CANIZARES, Massachusetts Institute of Technology, *Chair*
MARK R. ABBOTT, Oregon State University
FRAN BAGENAL, university of Colorado
DANIEL N. BAKER, University of Colorado
ROBERT E. CLELAND, University of Washington
GERARD W. ELVERUM, JR., TRW Space and Technology Group*
MARILYN L. FOGEL, Carnegie Institution of Washington
BILL GREEN, Former Member,U.S. House of Representatives
JOHN H. HOPPS, JR., Morehouse College
CHRIS J. JOHANNSEN, Purdue University
ANDREW H. KNOLL, Harvard University*
RICHARD G. KRON, University of Chicago
JONATHAN I. LUNINE, University of Arizona
ROBERTA BALSTAD MILLER, CIESIN-Columbia University
GARY J. OLSEN, University of Illinois, Urbana-Champaign
MARY JANE OSBORN, University of Connecticut Health Center
GEORGE A. PAULIKAS, The Aerospace Corporation
JOYCE E. PENNER, University of Michigan
THOMAS A. PRINCE, California Institute of Technology
PEDRO L. RUSTAN, JR., Ellipso, Inc.
GEORGE L. SISCOE, Boston University
EUGENE B. SKOLNIKOFF, Massachusetts Institute of Technology
MITCHELL SOGIN, Marine Biological Laboratory
NORMAN E. THAGARD, Florida State University
ALAN M. TITLE, Lockheed Martin Advanced Technology Center
RAYMOND VISKANTA, Purdue University
PETER VOORHEES, Northwestern University
JOHN A. WOOD, Harvard-Smithsonian Center for Astrophysics
JOSEPH K. ALEXANDER, Director

---

* Former member.

COMMISSION ON PHYSICAL SCIENCES, MATHEMATICS, AND APPLICATIONS

PETER M. BANKS, Veridian ERIM International, Inc., *Co-chair*
W. CARL LINEBERGER, University of Colorado, *Co-chair*
WILLIAM F. BALLHAUS, JR.,Lockheed Martin Corp.
SHIRLEY CHIANG, University of California at Davis
MARSHALL H. COHEN, California Institute of Technology
RONALD G. DOUGLAS, Texas A&M University
SAMUEL H. FULLER,Analog Devices, Inc.
JERRY P. GOLLUB,Haverford College
MICHAEL F. GOODCHILD,University of California at Santa Barbara
MARTHA P. HAYNES,Cornell University
WESLEY T. HUNTRESS, JR.,Carnegie Institution
CAROL M. JANTZEN,Westinghouse Savannah River Company
PAUL G. KAMINSKI,Technovation, Inc.
KENNETH H. KELLER,University of Minnesota
JOHN R. KREICK,Sanders, a Lockheed Martin Company (ret.)
MARSHA I. LESTER,University of Pennsylvania
DUSA M. MCDUFF,State University of New York at Stony Brook
JANET L. NORWOOD,Former U.S. Commissioner of Labor Statistics
M. ELISABETH PATÉ-CORNELL,Stanford University
NICHOLAS P. SAMIOS, Brookhaven National Laboratory
ROBERT J. SPINRAD,Xerox PARC (ret.)
NORMAN METZGER, Executive Director (through July 1999)
MYRON F. UMAN, Acting Executive Director

Foreword

Remote observations of Earth from space serve an extraordinarily broad range of purposes, resulting in extraordinary demands on those at the National Aeronautics and Space Administration (NASA), the National Oceanic and Atmospheric Administration (NOAA), and elsewhere who must decide how to execute them. In research, Earth observations promise large volumes of data to a variety of disciplines with differing needs for measurement type, simultaneity, continuity, and long-term instrument stability. Operational needs, such as weather forecasting, add a distinct set of requirements for continual and highly reliable monitoring of global conditions.

The present study confronts these diverse requirements and assesses how they might be met by small satellites. In the past, the preferred architecture for most NASA and NOAA missions was a single large spacecraft platform containing a sophisticated suite of instruments. But the recognition in other areas of space research that cost-effectiveness, flexibility, and robustness may be enhanced by using small spacecraft has raised questions about this philosophy of Earth observation. For example, NASA has already abandoned its original plan for a follow-on series of major platforms in its Earth Observing System.

This study finds that small spacecraft can play an important role in Earth observation programs, providing to this field some of the expected benefits that are normally associated with such programs, such as rapid development and lower individual mission cost. It also identifies some of the programmatic and technical challenges associated with a mission composed of small spacecraft, as well as reasons why more traditional, larger platforms might still be preferred. The reasonable conclusion is that a systems-level examination is required to determine the optimum architecture for a given scientific and/or operational objective. The implied new challenge is for NASA and NOAA to find intra- and interagency planning mechanisms that can achieve the most appropriate and cost-effective balance among their various requirements.

Claude R. Canizares, *Chair*
Space Studies Board

FOREWORD

Acknowledgment of Reviewers

This report has been reviewed by individuals chosen for their diverse perspectives and technical expertise, in accordance with procedures approved by the National Research Council's (NRC's) Report Review Committee. The purpose of this independent review is to provide candid and critical comments that will assist the authors and the NRC in making the published report as sound as possible and to ensure that the report meets institutional standards for objectivity, evidence, and responsiveness to the study charge. The contents of the review comments and draft manuscript remain confidential to protect the integrity of the deliberative process. We wish to thank the following individuals for their participation in the review of this report:

David Atlas, Atlas Concepts;
Peter Burr, consultant;
Greg H. Canavan, Los Alamos National Laboratory;
Leonard A. Fisk, University of Michigan;
Margaret G. Kivelson, University of California at Los Angeles;
Marlon R. Lewis, Dahlousie University, Canada; and
Eberhardt Recktin, consultant.

Although the individuals listed above have provided many constructive comments and suggestions, responsibility for the final content of this report rests solely with the authoring committee and the NRC.

ACKNOWLEDGMENT OF REVIEWERS

Contents

	EXECUTIVE SUMMARY	1
1	INTRODUCTION	7
	References	10
2	CORE OBSERVATIONAL NEEDS	11
	Required Measurements	11
	Characterization, Calibration, and Validation	16
	Data Continuity	17
	Simultaneity	19
	Sampling Errors	19
	Summary	20
	References	21
3	PAYLOAD SENSOR CHARACTERISTICS	22
	Payload Design and Accommodation Requirements	22
	Currently Planned Sensors	23
	Sensor Costs	26
	Future Sensor Designs: Implications of Advanced Technologies	26
	Summary	30
4	SMALL SATELLITE BUSES	31
	Capabilities of Small Satellite Buses	31
	Spacecraft Bus Costs	32
	Utility of "Commercial" Spacecraft	33
	Spacecraft Capability as a Payload Design Parameter	34
	Principal Investigator-led Projects	34
	Future Trends	35
	Summary	35
	References	36

5	SMALL LAUNCH VEHICLES	37
	Small Launch Vehicles for EOS and NPOESS	38
	Summary	40
6	SMALL SATELLITES AND MISSION ARCHITECTURES	41
	Options for Distributing Sensors	41
	Cost-effectiveness of Small Satellite Architectures	43
	Summary	48
	References	50
7	OPPORTUNITIES AND CHALLENGES IN MANAGING SMALL SATELLITE SYSTEMS	51
	Programmatic Approaches to Technical Issues	51
	Risks	52
	Summary	56
	References	57
8	FINDINGS AND RECOMMENDATIONS	58
	Mission Costs	58
	Meeting Mission Goals: Opportunities with Small Satellites	59
	Operational and Research Earth Observations	59
	Payloads	60
	Satellite Buses	61
	Launch Vehicles	61
	Mission Architectures	61
	System Management	62
	Summary	63
	APPENDIXES	
A	STATEMENT OF TASK	67
B	EFFECTS OF TECHNOLOGY ON SENSOR SIZE AND DESIGN	69
C	U.S. LAUNCH VEHICLES FOR SMALL SATELLITES	77
D	CASE STUDIES	82
E	ACRONYMS AND ABBREVIATIONS	90

Executive Summary

At the request of the National Aeronautics and Space Administration (NASA) and the National Oceanic and Atmospheric Administration (NOAA), the Committee on Earth Studies analyzed the capability of small satellites to satisfy core observational needs in Earth observing and environmental monitoring programs. The committee's study focused in particular on the use of small satellites to be inserted in the NASA Earth Observing System (EOS) program and the planned NOAA-Department of Defense (DOD) National Polar-orbiting Operational Environmental Satellite System (NPOESS) program.[1]

The committee's study was begun in November 1995, during a period of much debate over the feasibility and merits of substituting smaller satellites for larger systems. Proponents of the small satellite approach believed that advances in miniaturization would allow development of much smaller sensors with performance sufficient for many Earth science and operational needs. These smaller sensors could be accommodated on capable, smaller spacecraft and launched with the new generation of smaller launch vehicles. Further, they argued, performing missions with smaller payloads, spacecraft, and launch vehicles would lead to lower costs, greater programmatic flexibility, more and faster missions, and accelerated infusion of new technologies. These features would help fill recognized gaps and provide new opportunities in the nation's Earth observation programs.

The committee approached the study by setting out to understand the observational needs for key NASA and NOAA Earth remote sensing programs, and to determine and assess the availability and capability of sensors, satellite buses, and launch vehicles suitable for small satellite missions. The committee examined opportunities presented by small satellite options with respect to mission architecture and assessed their implications for future NASA and NOAA missions.

SMALL SATELLITES VERSUS SMALL MISSIONS

The committee found that, in addressing the role of small satellite missions, it is important to distinguish between small satellites, small missions, and larger missions employing small satellites. In this study, the term

---

[1] EOS is the space-based component of NASA's Earth Science Enterprise (formerly known as the Mission to Planet Earth program). Currently, the Department of Commerce, through NOAA, supports the Polar-orbiting Operational Environmental Satellite weather satellite system, and DOD, through the Air Force, supports the Defense Meteorological Satellite Program weather satellite system. NPOESS will be supported by NOAA and DOD and will be managed by the Integrated Program Office staffed by NOAA, DOD, and NASA personnel.

"small satellites" refers to *size*—satellites in the 100 to 500 kg class capable of meeting NASA and NOAA Earth observation measurement requirements. The term "small mission" refers to *cost*—that is, a small mission is a comparatively low-cost mission. NASA's current Earth science strategy of performing a larger number of smaller missions (versus that planned in earlier conceptions of the EOS program) is predicated on the cost of each mission being relatively low. Although small satellites may help enable low-cost small missions, not all small satellite missions will be low cost. **Low costs result as much from the relative simplicity of the mission (or the preexistence of mission elements) as from the size of the satellite.**

The ability to achieve low costs when employing small satellites for larger missions is even more uncertain than when small satellites are employed for small missions. For example, performing a mission with a large constellation of small satellites to achieve a high sampling frequency may cost a great deal, even though the individual satellites may cost little. A more controversial example is to use small satellites as a substitute for larger satellites to accommodate a specified complement of sensors. In this trade-off, the cost of initially placing the sensors into orbit may be higher with multiple small satellites because it involves building and launching more satellites. The lowest cost architecture to maintain a functioning complement of sensors over a prescribed mission lifetime depends on the system availability requirements (i.e., the percentage of time the system must be able to deliver the specified data) and the design life and reliability of the mission elements (sensors, spacecraft bus, launch vehicles).

MEETING CORE OBSERVATIONAL NEEDS

NASA's and NOAA's core Earth observational needs span many disciplines, including oceanography, land processes, atmospheric sciences, meteorology, climate, and geodesy. While these aspects of Earth studies have shared remote-sensing spacecraft, the mission goals for the different disciplines often have different mission time horizons, different orbit requirements, and differing instrument sizes and require measurements of differing resolution, repeat cycle, and area coverage, for example. Although it is sometimes necessary (or at least very desirable) that some of these data be temporally and geographically coincident to some tolerance, accommodating these diverse mission goals with large, multisensor spacecraft generally involves compromises. The committee has sought to understand these requirements and compromises to help assess the capabilities and opportunities associated with small satellites.

A primary argument for a multisensor platform is a requirement for temporal or spatial simultaneity of data collection—for example, when studying the interaction between columnar water vapor and temperature, or when there is a desire to test for the presence of clouds in the field of view. **However, the committee found the requirement for simultaneity difficult to prove.** Generally, only a need to observe clouds or other rapidly changing conditions supported the argument for simultaneity. **Rather, it is more important to ensure that a full suite of sensors is contemporaneously available** to measure processes related to coupling of various components of the Earth system, such as air/sea fluxes, and that this suite is continued for a sufficient period of time. For operational systems, strict simultaneity is also not generally required. Because the sensors are not all co-boresighted and because some have inherently different sampling strategies, even operational satellite platforms that carry multiple sensors mostly provide contemporaneous rather than simultaneous observations. Even in those cases where simultaneity *is* required, there may be opportunities to use alternative architectures—for example, clusters of satellites flying in formation.

Although there are differences between the operational measurement requirements of missions such as NPOESS and the Earth science research requirements of missions developed by NASA's Earth Science Enterprise, there is clearly overlap as well. Moreover, many operational measurements are useful for research, especially for long-term climate studies. The separation of instrument variability from the often subtle long-term variations in climate-related processes requires careful calibration and validation of the sensor and its derived data products. **As sensors are replaced over time, it is essential to maintain continuity of the data product despite changes in sensor performance ("dynamic continuity").**

The requirements for research missions evolve rapidly with advances in science and technology. Long development times associated with large multisensor missions often run counter to this emphasis on flying the latest in sensor design. Research missions emphasize the quality of the individual observation and thus constantly push the technology envelope in an attempt to obtain better-quality data. By contrast, operational systems tend to evolve more slowly, in part in response to budgets that grow more slowly and in part in response to the well-defined operational nature of the missions. For example, the data processing infrastructure of the user community often involves numerical models that may be expressly designed to assimilate satellite measurements collected at specific times with specific observing characteristics.

CAPABILITY OF SMALL SATELLITES TO PERFORM EARTH OBSERVATION MISSIONS

A review of development trends points to continued efforts to increase capability, reduce size, and lower costs of small satellite buses. In particular, **technology has advanced to the point where very capable buses are currently available for performing many Earth observation missions.** However, some Earth observation payloads are too large, too heavy, too demanding of power, or generate too much vibration to be accommodated efficiently with small satellite missions. Future advances in payload technology should mitigate this situation, but there are fundamental laws of physics that in some cases restrict the degree of miniaturization that can be achieved while retaining sufficient performance to meet the observation requirements. **Thus, the committee sees small satellites as a complement to larger satellites, not a replacement for them.**

FLEXIBILITY AND NEW OPPORTUNITIES PROVIDED BY SMALL SATELLITES

Small satellites offer new opportunities to address the core observational requirements of both operational and research missions. **Small satellites, in particular single-sensor platforms, provide great architectural and programmatic flexibility.** They offer attractive features with respect to design (distribution of functions between sensor and bus); observing strategy (tailored orbits, clusters, constellations); faster "time to science" for new sensors; rapid technology infusion; replenishment of individual failed sensors; and robustness with regard to budget and schedule uncertainties. New approaches to observation and calibration may be possible using spacecraft agility in lieu of sensor mechanisms, for example. Small satellite clusters or constellations can provide new sampling strategies that may more accurately resolve temporal and spatial variability of Earth system processes.[2] With advances in technology and scientific understanding, new missions can be developed and launched without waiting for accommodation on a multisensor platform that may require a longer development time.

Small satellite missions, as a new element of measurement strategy, may also help provide more balance between long-term operational or systematic observations and short-term experimental process measurements, as well as between focused missions and larger, more comprehensive missions. Programs can be more readily tailored to fiscal funding constraints when implemented as a series of smaller satellites (although this raises the risk of an incomplete data set unless the missions are planned and executed carefully).

AVAILABILITY OF RELIABLE LAUNCH VEHICLES

Achieving the full promise of small satellites will require the availability of reliable U.S. launch vehicles with a full range of performance capabilities. This is currently not the case. Present launch vehicle performance capabilities do not effectively span the range of potential payloads. For example, there is a significant gap in capability between the Pegasus-Athena-Taurus launch vehicles and the Delta II. Also, fairing volume (which determines the stowed payload size as well as the type and complexity of deployable systems such as antennas) is often limited and sometimes drives the size of the payload. More flexible launch systems are needed where volume constraints are less stringent. Further, early experience with the new small launch vehicles has included a number of failures, and the present paucity of reliable options is of great concern. This is likely due in part to the relative newness of these systems and a desire to minimize development costs for these commercial ventures. Continued development should overcome the difficulties and yield a suitable balance between cost and reliability.

---

[2] *Clusters* are a collection of two or more satellites relatively closely spaced in a common orbit (formation flying). *Constellations* (e.g., Global Positioning System, Iridium) are a collection of satellites whose relative positions are controlled in each of multiple orbits.

It will take some time, and likely some additional failures, before any of these launch vehicles establish a reliability record approaching that of the Delta II. Plans by numerous suppliers to address these needs are encouraging.

COST OF SMALL SATELLITE MISSIONS

Small spacecraft do offer opportunities for low-cost missions, but very low costs are experienced only with simple spacecraft performing limited missions. Small spacecraft can be relatively expensive when they retain the complexity required to meet demanding science objectives (pointing accuracy, power, processor speed, redundancy, etc.). Commercial "production" satellite buses offer the potential for reducing costs.[3] However, they generally have to be tailored—with attendant costs—to accommodate existing Earth observation payloads. Designing new payloads to match existing bus capabilities offers greater cost-effectiveness, but caution must be exercised not to compromise the scientific mission in so doing.

Several small missions—e.g., Clementine, QuikSCAT (Quick Scatterometer)—consisting of a single small satellite launched on one of the new class of small launch vehicles have been successfully performed at a relatively low cost. But the true cost of these missions is somewhat controversial in that they employed preexisting sensors and technology developed under separate funds. The true cost of a mission must also include the investment in technologies around which the activity is built. Leveraging advanced technology to lower mission costs is laudable, but understanding the true cost of the mission requires consideration of such prior investments, particularly when they are directly supportive of the mission (e.g., preexisting sensors).

SENSOR DEVELOPMENT

The factors driving mission development time and cost for Earth observation missions are typically associated with the development of sensors. "Standard" small satellite buses and launch vehicles are available to support faster missions, but development of new sensors will often control a program's schedule regardless of satellite size. Small satellites can provide a quicker path to operation and data collection if the required instruments—sized to the smaller spacecraft—are available, or if they are under development on a schedule that matches the development timetable of the spacecraft. **Many of the early successes with smaller, faster missions depended on the availability of sensors developed elsewhere (e.g., Clementine, QuikSCAT). Whereas larger mission budgets and schedules have traditionally provided for their own sensor development needs, continued success with fast, cheap, small missions will require a reservoir of new sensor technology developed through alternative sources.** If small missions are burdened with the development of their sensors, then the cost, the development time, and the time to science will increase accordingly.

MISSION ARCHITECTURE

The development of highly capable small satellites has given new flexibility to planners when designing mission architectures. Small satellites offer program managers flexibility that is useful for both operational and research missions. For example, operational missions might employ small satellites to ensure minimum gaps in critical data records, while research missions might use small satellites to ensure short time to science. Constellations or clusters of small satellites also afford new strategies for acquiring data or for accommodating fiscal funding constraints.

Larger, multisensor platforms have advantages as well. When needed, they provide a more stable platform and facilitate spatial and temporal simultaneity of measurements. Because fewer spacecraft and launches are involved, multisensor platforms offer a higher probability of placing a given complement of sensors into orbit without loss—and, often, at lower initial cost. Multisensor platforms frequently offer the simplest ground segment solutions, including mission operations, downlink and data system architectures, and calibration and validation of sensors.

---

[3] Commercial spacecraft buses are those for which there exists an operating production line serving a commercial market, as is the case for some communication satellites (e.g., Iridium).

The trade-off between small and large platforms is a complicated function of overall mission objectives, available budgets, tolerance for risk, and success criteria. These criteria are significantly different for research and operational missions. For example, operational systems are judged by performance, life cycle cost, and availability (the percentage of time the system can provide timely delivery of data). Loss of a single critical sensor can result in mission failure. Multiple launches of small satellites carry a higher risk of a launch or satellite failure, although the impact of such a failure with a larger multisensor satellite can be greater. Research missions are more tolerant to partial failure and place higher value on dynamic continuity and data quality as well as the flexibility to pursue new sensors and new science requirements aggressively.

The committee found that life cycle cost trade-offs between multisensor platforms and multisatellite architectures are driven by the reliability and design lives of the system elements (sensors, satellite buses, launch vehicles, ground segments) and by availability requirements for operational systems. The following conclusions pertain, depending on these requirements and system element characteristics:

- **The lowest cost to place a given set of sensors into orbit will often be with the smallest suitable multisensor platform.**
- **The lowest cost architecture to maintain a set of operational sensors in orbit for a sustained mission life is mission specific and must be determined on a case-by-case basis.**
- **Small satellites may provide economic benefits as part of a replacement strategy for failed sensors or for sensors with limited design life or reliability.**

Small and large satellite architectures show differing life cycle cost sensitivities to sensor reliability for sustained missions. As a result, there are conditions for which large satellite architectures are most cost-effective, as well as conditions that favor small satellites. Large satellite architecture costs are more sensitive to sensor reliability because larger satellites carry more sensors, all of which are replenished if a new satellite is launched in response to a critical sensor failure. When sensor reliability is high and failure infrequent, the lower cost of deploying the payload on fewer, larger platforms outweighs the added costs of occasionally launching unnecessary sensors and provides a life cycle cost advantage to large satellite architectures. But low sensor reliability, with concomitant frequent replenishment, leads to excessive unnecessary sensor replacement with large platforms, thus favoring small satellite architectures.

The often complex evaluation of whether the use of a small satellite is appropriate is driven by mission-specific requirements, including those related to the policy and execution of the program, fiscal constraints, and the scientific needs of the end users. Considering the many issues involved, **the design of an overall mission architecture, whether for operational or research needs, requires a complete risk-benefit assessment for each particular mission.** For some missions, a mixed fleet of small and large satellites may provide the most flexibility and robustness, but the exact nature of this mix will depend on mission requirements.

MANAGEMENT OF SMALL SATELLITE PROGRAMS

Innovative management approaches are needed to exploit the potential advantages offered by the small satellite approach if, as the committee believes, missions are to be science-driven versus technology-driven. New management approaches would benefit the development and implementation of calibration and validation strategies that maintain data continuity between sensors on successive satellites.

Fresh management approaches include streamlining program management and reducing management overhead, which can easily slow system development or discourage innovation, thus inhibiting many of the advantages of the small satellite approach. Small, tightly integrated teams have an advantage in such a development process as overhead costs decrease with team size. Experience shows, however, that efforts to reduce costs may result in severe pressures on the team. New approaches to program management can mitigate this problem. For example, government insight as part of the development team can limit the need for oversight by limiting formal reviews and documentation to those that truly add value.

Experience to date with small satellite missions offers many lessons on efficiencies achieved and risks associated with streamlined management techniques. Several missions have been quite successful, but delayed sensors, spacecraft development problems, launch vehicle availability and failure, and inadequate mission operations plans have all led to delays, cost increases, cancellation, and/or total loss. We must learn from these successes and failures to attain the full promise of small satellites in the future.

A common theme from the cases studied is that **the attempt to achieve faster and cheaper missions by streamlining operations and reducing non-value-added tasks must also include plans to maintain balance among all program elements. Imbalances among the sensor, spacecraft bus, launch vehicle, and ground system elements can lead to serious inefficiencies and risks.** Risk must be carefully assessed for all program elements when defining the system, particularly for schedule-critical missions. For the greatest cost-effectiveness, risk should be continuously assessed, progress monitored, and plans adjusted to keep the total program in balance. There is also a need for well-defined, well-understood, and consistent roles for government and industry partners and regular communication between all parts of the team.

MISSION PLANNING

User tolerance of risk is a key consideration when planning research or operational Earth observation programs. Some Earth science missions require access to long-term, consistent data sets from a variety of sensors. Operational systems, such as meteorological satellites, have strict requirements for data availability from multiple sensors for short-term and long-term forecasting. Although the risks for the individual small satellite components may be higher, small satellites may allow the design of a resilient, robust system (e.g., constellation of satellites) where the total mission risk is smaller. **Thus, management structures must not only allow the benefits of small satellites to be realized, but must also enable assessment and mitigation of the new set of risks posed by new mission architectures.**

Traditional procedures to develop mission and sensor concepts and the associated peer review process need to be streamlined. First, there must be appropriate mechanisms to ensure the design and maintenance of a coherent observing strategy. For example, solicitations for new NASA science missions should be consistent with the overall science directions of the Earth Science Enterprise. Second, management must address the issues associated with maintaining dynamic continuity of long-term data sets where the specific sensors (and even measurement techniques) will change over time. A comprehensive plan for cross-sensor calibration, data validation, and pre-launch characterization is especially important for climate research. Third, the science community must be prepared to make quantitative evaluations of sampling issues versus measurement quality in regard to the overall quality of the data products. This includes an evaluation of the impacts of data gaps as well as of levels of temporal and spatial resolution. The science community should be involved throughout the system design and implementation process rather than be limited to providing measurement requirements at the initial design stages. Regular assessments of sensor and system design, data products, and algorithms are needed to provide science community insight into the process.

CONCLUSION

The committee finds that the maturation of remote sensing science and the development of new sensor, platform, and launcher technologies now allow a more flexible approach to both research and operational Earth remote sensing. Small satellite missions have provided and should continue to provide an important component of how Earth observations are conducted from space. However, their limitations—both evident and more subtle—suggest that they are not an appropriate substitute for all larger satellites. **The committee recommends that, in planning for future NASA and NOAA missions, the choice of mission architecture should be driven by the mission requirements and success criteria, and an optimum solution should be sought, whether with large, mid-size, small, or a mixed fleet of platforms. The committee also recommends that both the research and operational communities perform a complete analysis of sampling strategies in the context of potential new mission architectures.**

1

Introduction

In November 1995, the Committee on Earth Studies of the National Research Council's Space Studies Board began a study to analyze the capability of small satellites[1] to satisfy core observational needs in Earth observation and weather monitoring programs of the National Aeronautics and Space Administration (NASA) and the National Oceanic and Atmospheric Administration (NOAA). NASA's interest in the possible use of smaller satellites for its Earth Observing System (EOS)[2]—the space-based component of the Office of Earth Science's Earth Science Enterprise—arose from budgetary pressures, the desire within the scientific community for more missions and shorter mission time lines, the perception that there are both gaps and unrealized opportunities in the enterprise's space segment, and the determination of NASA officials to accelerate the infusion of new technologies into their space programs.

The Department of Commerce (specifically NOAA) and the Department of Defense (DOD, specifically the Air Force) have also been examining potential roles for small spacecraft[3] as they proceed with plans to develop a converged polar-orbiting weather satellite system, NPOESS (National Polar-orbiting Operational Environmental Satellite System), scheduled for launch in approximately 2009 (availability date 2008). Prior to the start of the NPOESS program, both agencies had been planning block changes and upgrades to their existing polar-orbiting weather satellites, Polar-orbiting Operational Environmental Satellites (POES) and the Defense Meteorological Satellite Program.[4] The NPOESS program is being executed by the Integrated Program Office with representatives from DOD, NOAA, and NASA. NASA's particular interest in NPOESS involves its potential to fulfill long

---

[1] There is no generally accepted definition of a small satellite. In the literature, the term covers everything from a satellite weighing from a few to 1,000 kg. The committee is primarily interested here in that class of satellites with sufficient capability for application in NASA and NOAA Earth observation programs. Such satellites, which tend to cluster in the 100 to 500 kg range, can provide robust payload accommodations for one or more instruments and are suitable for launch on the new class of low-cost, small expendable launch vehicles or as part of a multisatellite launch on a larger expendable launch vehicle. They represent scheduled missions rather than "flights of opportunity," where the satellite is piggybacked on another mission.

[2] The EOS program is "the centerpiece of NASA's Earth Science Enterprise. It consists of a science component and a data system supporting a coordinated series of polar-orbiting and low-inclination satellites for long-term global observations of the land surface, biosphere, solid Earth, atmosphere, and oceans" (EOS Project Science Office, 1999). Until January 1998, the Office of Earth Science and the Earth Science Enterprise were called the Office of Mission to Planet Earth and the Mission to Planet Earth program, respectively.

[3] The terms "satellite" and "spacecraft" are used interchangeably in this report.

[4] The National Performance Review and Presidential Decision Directive NSTC-2 (May 1994) directed the Departments of Defense and Commerce and NASA to establish a converged national weather satellite program. This program, NPOESS, combines the follow-on to the

term systematic measurement requirements formerly planned for follow-on missions to the first series of EOS satellites, especially the PM[5] series because of its focus on monitoring weather- and climate-related variables. Each NPOESS satellite is currently planned as a multisensor spacecraft. However, alternative system architectures are possible that would distribute the sensor complement among a larger number of smaller satellites.

The committee's study originated during a period when satellite builders and policymakers were engaged in a spirited debate over the feasibility and merits of substituting smaller satellites for larger systems. Advances in miniaturization were said to allow much smaller sensors that retained sufficient performance for many Earth science and operational needs. These smaller sensors could be accommodated on smaller spacecraft, which would be smaller still because of miniaturization of various spacecraft subsystems. Reducing the size, volume, and weight of both payload and spacecraft would then allow the use of either the new generation of smaller launch vehicles or clustering of spacecraft on a single launch of a larger launch vehicle. It was argued that performing missions with smaller payloads, spacecraft, and launch vehicles would lead to dramatically lower costs.

The debate over the use of small satellites had sometimes been portrayed as a dispute between innovative satellite designers and government bureaucrats or industry officials who either lacked vision or had financial incentives to maintain the status quo. The committee found that these characterizations were either inaccurate or a simplification of more complex circumstances. It is noteworthy, for example, that the historic providers of large Earth remote sensing satellites have also provided small satellite systems for space physics research, planetary exploration, and other space missions. In addition, it was evident to the committee that any credible discussion of small versus large had to include a detailed analysis of the many interrelated technical and programmatic issues associated with the design and development of satellite systems.

In responding to its charge (Appendix A), the committee set out to understand the observational needs for key NASA and NOAA Earth remote sensing programs and to determine and assess the availability and capability of sensors, satellite buses, and launch vehicles suitable for small satellite missions. Further, the committee examined opportunities presented by small satellite options with respect to mission architecture and assessed their implications for future NASA and NOAA missions.

During the study, both NASA and NOAA made programmatic decisions that affected the committee's course. NASA restructured its Earth science program such that missions that would follow the initial EOS AM, PM, and Chemistry satellites would be smaller, more flexible, and responsive to advances in technology and science. NASA also planned to integrate EOS missions with operational weather satellite programs (e.g., NPOESS) for long-term systematic measurements. Further, the NOAA-DOD-NASA Integrated Program Office opted to develop new sensors, as opposed to continuing with heritage EOS sensors, for critical NPOESS measurements through competitive procurements. Thus, both NASA and NOAA plans now recognize and embrace current capabilities and ongoing advances in sensor and spacecraft technology for future Earth observation missions. Consequently, the committee altered its planned response to its charge and de-emphasized the study of specific new technologies in favor of an increased emphasis on the implications and impact of capable small sensors and satellites on mission architecture and management trade-offs. Among the questions emphasized in this modified approach were these:

- Are there sustained opportunities for low-cost, quick-response, focused missions, leading to a reduced "time to science" (analogous to the commercial sector's "time to market")?
- Would affordable constellations of small satellites open the door to enhanced science via more frequent or continuous sampling strategies?

Defense Meteorological Satellite Program and the POES program. An integrated tri-agency office was established on October 1, 1994, to manage acquisition and operations of the converged satellite program.

[5] The EOS satellites will be launched into polar, Sun-synchronous orbits. The EOS PM satellite will cross the equator at 1330 local time. The EOS AM satellite will cross the equator at 1030 local time. The afternoon and morning crossings facilitate observations of atmospheric and land processes, respectively.

The committee explored the scientific merits and technical capabilities of small satellites; the development status (e.g., availability and reliability) of the necessary system elements; and the programmatic aspects of implementing small satellite missions. The criteria used to assess small satellite utility and to examine mission architecture trade-offs included performance capability, risk, mission flexibility and robustness, and the potential for streamlined management processes—all with a focus on the potential for lower mission-life-cycle costs. Several case studies (Appendix D) were examined to assess the reality versus the promise of small satellites to date and to help identify paths to greater success in the future.

Both operational and research programs were considered, and the distinction between them underlies much of the discussion in this report. Characteristics of operational programs include an established community of data users who depend on a steady or continuous flow of data products, long-term stability in funding and management, a conservative philosophy toward the introduction of new technology, and stable data-reduction algorithms. Research programs often require greater measurement accuracy, more attention to calibration, programmatic flexibility, and faster time to science; depending on the cost of the mission, they can be more tolerant of risk.

The study emphasized the launch and space segments of Earth observation missions. Although treated more superficially in this study, ground segment operations (communications, command and control, and data routing and processing) and space system infrastructure (ground and space assets) may weigh heavily on mission architectures that involve many satellites—and may merit a study of their own.

During the course of its work, the committee heard presentations from companies long involved in producing small satellites for both commercial and research use. The committee also heard from industry representatives involved with NASA's Small Spacecraft Technology Initiative (SSTI),[6] a very aggressive program initiated by NASA in 1994 that attempted to demonstrate faster, better, and cheaper approaches to the development of small satellites. Key questions addressed to all study participants were whether the use of smaller satellites could reduce overall mission cost and what the controlling factors were.

In considering the potential for small satellites to reduce the cost of Earth observation missions, it is important to distinguish between small satellites, small missions, and larger missions employing small satellites. As noted in footnote 1, the present study's analysis of small satellites refers to those in the 100 to 500 kg class carrying one or a few sensors that are capable of acquiring data of the kind and quality required by NASA and NOAA for their EOS and POES/NPOESS programs. In this report, the term "small mission" refers to a comparatively low-cost mission. NASA's current Earth science strategy of performing a larger number of smaller missions (versus that planned in the early 1990s) is predicated on the cost of each mission being relatively low.[7] Low-cost satellite buses help enable low-cost missions, but total mission costs include those for the satellite (i.e., payload plus bus), launch vehicle, and mission operations. A number of small missions consisting of a single small satellite launched on one of the new class of small launch vehicles have been successfully performed at relatively low total cost,[8] albeit at a high specific cost—i.e., cost per pound (see Chapters 4 and 5). These missions take advantage of small

---

[6] SSTI was developed by NASA's Office of Space Access and Technology to advance the state of technology and reduce the costs associated with the design, integration, launch, and operation of small satellites (NASA SSTI, 1994). TRW and CTA Space Systems were each awarded a contract by NASA to design and launch small Earth observing satellites, which were subsequently named Lewis and Clark, respectively. The Lewis spacecraft was successfully launched from Vandenburg Air Force Base into its initial orbit on August 22, 1997. However, on August 26, 1997, an in-flight anomaly led to loss of attitude control and a discharged battery, which resulted in the eventual loss of the mission.

On February 25, 1998, NASA issued a press release announcing the termination of the Clark Earth science mission. The mission was terminated after an investment of some $55 million "due to mission costs, launch schedule delays, and concerns over the on-orbit capabilities the mission might provide." NASA retained Clark's launch vehicle services. See Steitz (1998). Appendix D of this report provides further discussion of the Lewis and Clark missions.

[7] Small mission costs are typically constrained by a limit prescribed in the Announcement of Opportunity (e.g., the Earth System Science Pathfinder mission cost was capped at $90 million in its 1996 Announcement of Opportunity).

[8] The true cost of a mission must also include the investment in technologies around which the activity is built. When ample advanced technology development has been done with prior investment that can be leveraged by a mission, the development costs of the mission itself may appear small. Discussions of the true cost of the mission should acknowledge such prior investments, particularly when they are directly supportive of the mission (e.g., preexisting sensors).

satellite buses that are available at about 15 to 50 percent of the cost of their larger Delta or Atlas class counterparts, depending on the difficulty of the mission requirements.

Mission cost trends are more uncertain when using small satellites to perform larger missions. For example, a mission architecture that employs a constellation of small satellites to achieve a high sampling frequency may cost a great deal, even though the individual satellites may cost little. More controversial is a mission architecture that accommodates a specified complement of sensors with several small satellites rather than with a larger multisensor satellite. In this trade-off, there is no a priori right answer on relative mission architecture costs as they depend on many variables (see Chapter 6).

The chapters that follow address first NASA's and NOAA's core observational needs and then three specific aspects of flight missions: sensor payloads, satellite buses, and launch vehicles. Finally, the report examines a number of systemwide issues, first with respect to overall mission architecture and then regarding several key management concerns that go beyond hardware development considerations. Specifically, Chapter 2 provides an overview of the measurements planned by NASA and NOAA to support satellite-based research and operational Earth observation programs, and it introduces key issues common to the development of either large or small satellite programs to fulfill NASA and NOAA requirements for the EOS and NPOESS programs. Chapter 3 provides a tutorial on the principles guiding the design and accommodation of sensors on a satellite. It also presents a discussion of sensor costs and an overview of the trade-offs and physical limits that govern sensor design. Chapter 4 discusses the capabilities of small satellite buses and their suitability for performing Earth observing missions. It also addresses some of the issues and trade-offs related to acquisition and cost, including the use of commercial, standard, and catalog buses. Chapter 5 addresses the current dilemma regarding the fact that achieving the full promise of small satellites will require the availability of reliable U.S. launch vehicles with a complete range of performance capabilities.

Chapter 6 is a key chapter in this report. Small satellites on dedicated launch vehicles offer a very high degree of programmatic flexibility, which allows them to be included in system trade-off studies that analyze the cost and effectiveness of alternative mission architectures for current and future programs. These trade-offs are illustrated in an analysis of alternative mission architectures for the NPOESS mission. Chapter 7 examines issues related to the management of small satellite programs, including consideration of science-driven versus technology-driven approaches and of calibration and validation strategies.

Chapter 8 reviews the preceding chapters in the broad context of the overall study, providing an integrated summary of key findings and recommendations.

REFERENCES

Earth Observing System (EOS) Project Science Office. 1999. EOS homepage. Available online at <http://eospso.gsfc.nasa.gov/eospso_homepage.html>.
National Aeronautics and Space Administration, Small Spacecraft Technology Initiative (NASA SSTI). 1994. Fact sheet: Smallsat—A new class of satellite. Available online at <http://ranier.hq.nasa.gov/SSTI_Page/NewClassSat.html>.
Steitz, D. 1998. NASA terminates Clark Earth science mission. Press Release 98–35, February 25. Washington, D.C.: National Aeronautics and Space Administration.

2

Core Observational Needs

This chapter summarizes National Aeronautics and Space Administration (NASA) and National Oceanic and Atmospheric Administration (NOAA) observational requirements for both research and operational satellite programs. Much of this discussion is not specific to small satellites but is provided to frame information presented in later chapters.

Research programs such as the Earth Observing System (EOS) take advantage of satellites' ability to provide a consistent, global vantage point from which to observe land, ocean, and atmosphere processes. *Operational* programs such as the Polar-orbiting Operational Environmental Satellite (POES) program also rely on the global perspective of satellites, but they generally emphasize the rapid delivery of global observations to support weather forecasting. The distinction between research and operational programs has a profound impact on how NASA and NOAA manage their programs. The distinction between research data sets and operational data sets can sometimes be artificial, however, and both can be essential elements of a climate research and global change program.

Operational systems are usually associated with the acquisition of long time series of data that may not meet the measurement requirements for climate research. An example of such a data set might be ocean topography, which appears in both the EOS and the National Polar-orbiting Operational Environmental Satellite System (NPOESS) observation sets. In this case, the accuracy requirements for climate research needs are much more stringent than the operational needs. On the other hand, some operational data sets do meet climate research needs, such as the NOAA POES and Geostationary Operational Environmental Satellites (GOES).

In the next section, the committee examines measurements identified by the EOS and NPOESS program offices as critical to the success of their respective missions. These measurements are not examined individually in detail; instead, the focus is on identifying common attributes present in each class of variable (research and operational).

REQUIRED MEASUREMENTS

Measurements in Support of Climate and Global Change Research

The specific variables that are currently measured from space are based on an understanding of Earth processes as well as on the technical ability to make measurements with the requisite accuracy and temporal and spatial resolution. The development of measurement requirements—and the generation of any list of critical measurements—is thus in continuous flux as scientific understanding evolves. For any particular mission, science requirements must be translated into a set of satellite sensors with specific measurement and sampling capabilities. The actual sensor requirements are therefore a melding of these science requirements and existing capabilities.

Many critical processes do not have an electromagnetic signal that can be measured by satellite. For example, the partial pressure of CO_2 in the surface ocean cannot be measured remotely, although it plays a critical role in determining the flux of CO_2 between the ocean and atmosphere. Also, many processes simply cannot be measured with adequate temporal and spatial resolution from space. For example, ocean salinity can be measured by satellite, but not with the required accuracy or spatial resolution of current microwave radiometer technology. Another example is the study of the ozone hole. In this case, ground observations first revealed the existence of the hole, which then stimulated a reanalysis of the satellite data sets. However, ground-based and in situ observations continued to be required to study the dynamics of the Antarctic ozone vortex in conjunction with satellite measurements. As these examples all show, Earth science measurement requirements are tempered by the reality of the technical capabilities of present and planned remote sensing systems.

The objective of this section is to identify the processes that are used to develop Earth science requirements and how these are in turn used to define a satellite mission. In this regard, the suite of 24 EOS measurements[1] shown in Box 2.1 represents the current understanding of the important processes related to Earth's climate and global changes as well as the ability of EOS sensors to make these measurements.[2] However, even when a measurement is listed, it should not be assumed that it will meet the science requirement. This is a result of the gaps in our understanding of Earth system processes, not poor sensor design. With this in mind, the EOS requirements were designed to be broad in scope, with the expectation that new insights into climate and global change processes will arise from having long-term, consistent observations. The EOS measurement set was also based on the realization that multiple observations of the same variable would lead to a better understanding of the relevant processes. That is, each observation has its own sampling characteristics and measurement approach that, when combined with other measurements of the same variable, may lead to a higher quality measurement.

In Earth system research, it is necessary to balance long-term observations with the need to study smaller scale events. Of particular interest are variations in the Earth system that occur on interannual and longer time scales. It will take many years to decades to observe such processes in a statistically robust manner. Processes that occur on much shorter time scales, such as severe storms or mesoscale ocean eddies, may drive the overall system, however. The Earth does not operate as a smoothly varying system but rather as a set of nonlinear processes that can change rapidly.

Earth system research goes far beyond the realm of atmospheric dynamics. The ocean clearly provides strong feedback through the transport of heat and the exchange of water with the atmosphere. Moreover, both the marine and terrestrial components of the biosphere affect climate through their impacts on heat and moisture exchanges as well as through their modulation of biogeochemistry, especially greenhouse gases. In other words, there is no single measurement that will provide a comprehensive understanding of climate processes and their interaction with the biosphere. Any Earth observing system must consist of an integrated, comprehensive set of measurements. However, it must also have the capacity to include new measurements as our understanding of the Earth system evolves and our technical abilities improve.

Measurements in Support of Operational Applications

The measurement requirements for operational observing systems, such as NPOESS, are designed for a set of objectives that differ from those for research observing systems. In large part, this is a result of operational systems usually being focused on short-term, event-scale processes and the rapid delivery of near-real-time data. Such applications place less importance on long-term stability of data sets and more importance on data availability, for example, to protect life and property. Operational data also play an important role in numerical weather prediction

---

[1] It is expected that the 24 EOS measurements discussed here—a set maintained during several previous program rescopings—will now change as NASA rethinks its plans beyond the first series of EOS spacecraft in light of this study's findings.

[2] See the sections on calibration and validation later in this chapter.

models that rely on data assimilation. Here, data and sampling needs are well understood from analyses of model performance. In fact, the models are often designed for specific data types with specific characteristics, so it is difficult to adapt them to assimilate new data sets. The primary user community is also well defined, and sensors and data products are often developed to meet particular application needs.

BOX 2.1 CLASSIFICATION OF THE EOS OBSERVATION SET INTO 24 MEASUREMENT CATEGORIES

Atmosphere
Cloud properties, including amount, optical properties, and height
Radiative fluxes at the top of the atmosphere and the surface
Precipitation
Tropospheric chemistry, including ozone and precursor gases
Stratospheric chemistry, including ozone, BrO, OH, and trace gases
Aerosol properties in both the troposphere and stratosphere
Atmospheric temperature
Atmospheric humidity
Lightning, including events, area, and flash structure

Solar Radiation
Total solar irradiance
Ultraviolet spectral irradiance

Land
Land cover and land use change
Vegetation dynamics
Surface temperature
Fire occurrence, including extent and thermal anomalies
Volcanic effects, including frequency of events, thermal anomalies, and impact
Surface wetness

Ocean
Surface temperature
Phytoplankton and dissolved organic matter
Surface wind fields
Ocean surface topography, including height, waves, and sea level

Cryosphere
Land ice, including ice sheet topography, ice sheet volume change, and glacier extent
Sea ice, including extent, concentration, motion, and temperature
Snow cover, including extent and water equivalence

There is, however, a large group of users outside the primary user community of weather services: The success of the POES data in furthering understanding of such long-term phenomena as El Niño has considerably broadened the user base. Secondary data products are available from the National Weather Service (NWS), the National Environmental Satellite Data and Information Service, and the National Centers for Environmental Prediction/Climate Prediction Center, which continue to expand the interest in these systems; research use of the data from these systems is increasing. It is not clear, however, that this use has driven the requirements for NPOESS.

The net result of these requirements is a suite of satellite sensors that evolve slowly over time and that are designed to meet well-defined needs. Given the operational focus of these missions, the user community has a very low tolerance for gaps in the data. In contrast to the Earth science missions, the primary users of operational data tend to rely on recently collected data rather than analyses of historical data. Moreover, the emphasis on specific application needs generally results in less interest in complex sensor suites to study a wide variety of Earth system processes. This is not to say that the sensor suites are not complex, but rather that they are chosen for specific application to the meteorological problem. The weather services are not usually interested in more general sensors that have a wide application to other problems. Indeed, the aversion to risk that characterizes operational programs such as those of the NWS is reflected in the conservative choice of sensors.

The emerging NPOESS program has developed a list of environmental data records (EDRs) that are intended to meet the needs of its broadly based user community. These EDRs are described in detail in the Integrated

> **BOX 2.2 ENVIRONMENTAL DATA RECORDS IDENTIFIED FOR THE NATIONAL POLAR-ORBITING OPERATIONAL ENVIRONMENTAL SATELLITE SYSTEM**
>
> **Key Parameters[1]**
> Atmospheric vertical moisture profiles
> Atmospheric vertical temperature profiles
> Imagery
> Sea surface temperature
> Sea surface winds
> Soil moisture
>
> **Atmospheric Parameters**
> Aerosol optical thickness
> Aerosol particle size
> Ozone total column/profile
> Precipitable water
> Precipitation (type and rate)
> Pressure (surface and profile)
> Suspended matter
> Total water content
>
> **Cloud Parameters**
> Cloud base height
> Cloud cover and layers
> Cloud effective particle size
> Cloud ice water path
> Cloud liquid water
> Cloud optical depth and transmittance
> Cloud top height
> Cloud top pressure
> Cloud top temperature
>
> **Earth Radiation Budget Parameters**
> Surface albedo
> Downward long-wave radiation at the surface
> Insolation
> Net shortwave radiation at the top of the atmosphere
> Solar irradiance
> Total long-wave radiation at the top of the atmosphere
>
> **Land Parameters**
> Land surface temperature
> Normalized difference vegetation index
> Snow cover and depth
> Vegetation and surface type

Operational Requirements Document.[3] For example, some of the service branches of the Department of Defense (DOD, which is NOAA's partner in the NPOESS program) require constant ground resolution imagery. Providing images in this format eases their interpretation and eliminates the need for image analysts to undergo the specialized training they would otherwise need to interpret images in the format provided by the Advanced Very High Resolution Radiometer instrument on POES. However, it is much less useful for quantitative analysis for scientific purposes because of the difficulty in obtaining quantitative radiometric measurements. The current list of EDRs (Box 2.2) represents a careful balance between NOAA and DOD needs. Each EDR has an associated threshold and objective for performance. The EDR objectives constitute a set of desired performance levels but are not strict requirements. If a sensor cannot meet the EDR threshold, however, it is deemed to have failed. The NPOESS

---

[3] This and other NPOESS requirements documents are available online at <CS:WebLink>http://npoesslib.ipo.noaa.gov/Req_Docs.htm>.

Ocean and Water Parameters	Currents Freshwater ice motion Ice surface temperature Littoral sediment transport Net heat flux Ocean color and chlorophyll Ocean wave characteristics Sea ice age and motion Sea surface height and topography Surface wind stress Turbidity
Space Environmental Parameters	Auroral boundary Auroral energy deposition Auroral imagery Electric field Electron density profiles and ionospheric specification Geomagnetic field In situ ion drift velocity In situ plasma density In situ plasma fluctuations In situ plasma temperature Ionospheric scintillation Neutral density profile/neutral atmospheric specification Radiation belt and low-energy solar particles Solar and galactic cosmic ray particles Solar extreme ultraviolet flux Suprathermal through auroral energy particles Upper atmospheric airglow

Note: These were current as of this writing; the EDRs are, however, subject to change, and the latest requirements may be found a the NPOESS Web site at <CS:WebLink>http://npoesslib.ipo.noaa.gov>.

[1] These are baseline measurements that must be part of NPOESS.

requirements methodology may appear to be more rigorous than the open-ended approach taken with climate research, but note that each approach has been chosen to meet the needs of its particular user community.

Although there are apparent overlaps between the EOS 24 measurement set and the NPOESS EDRs, the lists are not interchangeable. Each measurement has its own set of performance and sampling requirements that are appropriate for a specific application. In general, sensor performance requirements for climate and global change research are more stringent than operational requirements, while the operational requirements for sampling and continuity are more stringent than the science requirements. However, as Earth system research matures, the operational observing systems probably will begin to assume responsibility for these more stringent measurement requirements. For example, the long-term variation in climate processes requires a monitoring approach that is appropriately done in an operational context. This transfer of measurement responsibility from research to operational systems must take place in an orderly manner.

CHARACTERIZATION, CALIBRATION, AND VALIDATION

Prelaunch Sensor Characterization

Prelaunch characterization is necessary for all sensors that are expected to return accurate data, such as those required for climate studies. Sensors require a suite of prelaunch tests to establish performance parameters to be used in data processing. These performance parameters are instrument characteristics such as spectral bandpass, polarization sensitivity, out-of-field-of-view responsivity (scattered light effects), deviations from linearity, and temperature sensitivity. The type of performance parameter, as well as the accuracy to which it needs to be measured, is determined by the requirements of the data processing algorithm(s); this is true for both climate and operational sensors. However, the two types of sensors will differ in the extent and degree of the characterization tests.

Sensor characterization is different from sensor calibration. Calibration is the process of measuring the relationship between sensor output (digital counts) and absolute radiant (radiance) input. Sensor characterizations must be completed before launch in almost all instances. However, it is possible to calibrate a sensor's response "vicariously" on orbit by observations of calibrated ground targets. It is difficult, if not impossible, to measure most sensor characteristics after launch. Sensor characterization measurements in the laboratory can be accomplished under reasonably controlled conditions compared to on-orbit measurements. It may be sufficient to perform some characterizations at the component level and others at the system level; for particularly critical performance parameters, both system and component level measurements may be required.

The requirement for a complete set of accurately known sensor characteristics is independent of satellite size. Thus, it is not a unique problem of small satellites but may be more difficult to accomplish for them because of the funding and schedule limitations of their missions. To complete the tasks associated with full characterization of a sensor takes time and money, which may not be commensurate with the shorter schedules and lower costs desired for small satellite missions.

Calibration

Careful sensor calibration is essential if the typically small signatures of phenomena like climate change are to be recognized. For example, small increases in global sea surface temperature represent an enormous change in the heat content of the world ocean, and they are a valuable diagnostic of climate models. However, drift in sensor calibration or sensor performance could easily mask these changes. Monitoring solar output is another example where careful calibration is necessary; it is complicated in this case by the need to assemble a consistent time series across multiple copies of a sensor, variations in which are a significant component of the total apparent system variance.[4]

Calibration programs must be an ongoing part of any satellite system that is focused on Earth system studies. Much of the effort of calibration occurs before the satellite is launched. As with sensor characterization, this too requires time and money and may not be commensurate with the shorter schedules and lower budgets desired for small satellite missions.[5] These programs will rely on a combination of in situ studies—vicarious calibration—as well as on-board calibration systems. The primary role of an on-board calibration system is to measure the sensor's precision—that is, its short-term stability on orbit and its calibration stability through the rigors of launch. On-board calibration systems take space and add cost. It has not yet been demonstrated that they can be implemented within the constraints of the "smallsat" paradigm.[6] Calibration programs must pay particular attention to assessing the performance of new sensors, even if they are copies of old sensors. This assessment may mean overlapping the use of old and new sensors so it will be possible to ensure compatibility between data products.

---

[4] See Willson (1997) and the references therein.

[5] The care, planning, and effort that may be required are illustrated by the program to calibrate the EOS AM-1 Multi-Angle Imaging Spectroradiometer instrument. See Bruegge et al. (1996, 1998).

[6] However, the excellent results obtained from SeaWiFS (Sea-Viewing Wide Field-of-View Sensor) are an example of what can be done and illustrate the value of on-board calibration for a small sensor.

Alternatively, carefully planned field campaigns may provide sufficient cross-comparisons to produce a consistent time series across multiple sensors.

The present series of polar and geosynchronous meteorological satellites (POES, Defense Meteorological Satellite Program, GOES) were not designed to meet the rigorous calibration requirements of users such as climate researchers. While not yet defined, the calibration plan for the next generation of meteorological satellites, NPOESS, is likely to be less rigorous than that required for studies of climate and other aspects of global change. The committee notes, however, that satisfying many of the NPOESS operational requirements will necessitate comparisons between present measurements and climatologies derived from long-term observations. As undetected shifts in sensor performance may substantially reduce the quality of the operational product, regular assessments of instrument performance should be part of the calibration program.

The importance of an adequate calibration program is illustrated by the following example. The blended sea surface temperature (SST) was a NOAA-generated data product of the early 1980s. The SST was a combination of satellite and in situ measurements. However, if the satellite observations exceeded the climatological SSTs by too large a value, it was assumed that they were in error. In that case, the satellite-derived SSTs were replaced by the climatologically derived SST values, as it was assumed that the on-board calibration information was insufficient to evaluate sensor performance. The result of this approach, however, was that the blended SST product did not reveal the 1982–1983 El Niño Southern Oscillation (ENSO) event until direct measurements were made by ship in the eastern tropical Pacific—6 months after the ENSO event began. The satellite values, which were much warmer than the climatology measurements, had been correct, but they had exceeded the quality threshold and were discarded.

Validation

Validation is the process of evaluating the algorithms that are used to convert instrument measurements into geophysical quantities and assessing the uncertainties in derived geophysical quantities. As with calibration, validation is required for both research and operational systems, although the requirements may differ. The need for validation applies to all remote sensing systems, regardless of mission or satellite size.

Algorithms may change over time as new methods are developed. However, operational agencies are typically not interested in reprocessing old data in an attempt to extract new information. This essential difference between climatology and meteorology affects mission goals; it must be understood if the operational and research communities are to exchange data and rely on each other's satellites and databases. Validation should be done over the full range of possible environmental conditions. The EOS program is developing a plan for validation that should provide critical information for future researchers as they assemble long time series of observations from a variety of sensors. Validation can be seen as the counterpart of calibration, but it is applied to the software or algorithm component of the data set.

DATA CONTINUITY

A stable set of well-calibrated, reliable measurements is central to the U.S. Global Change Research Program. Indeed, much effort has been expended to design instruments that will be stable enough to ensure that they indicate actual changes in the Earth and its environs, rather than the effects of instrument artifacts and instabilities. The introduction of new instruments can undermine continuity and confidence in long-term measurements. Yet there are serious problems in attempting to field a particular set of instruments in perpetuity. For example, changes in technology will eventually make it impossible to reproduce a given set of instruments as the availability of the specific components and intellectual skills associated with a particular design vanish. Twenty years is probably the outside limit for any one design. In fact, some space instruments have been in production for nearly this interval, but closer examination would reveal that they have been continually upgraded to surmount some of the parts availability problems.

Continuity of measurements is sometimes confused with continuity of sensors. Given budgetary as well as user-imposed constraints, continuity may be ensured by a strategy of launching an identical sensor upon antici pated failure of an orbiting sensor, although other approaches do exist. NPOESS and the operational user community obviously have much stricter requirements for continuity, especially for the six key parameters listed in Box 2.2.

Operational Data Continuity

Operational data requirements, such as those of the NWS, are often based on the need to protect lives and property; therefore, a long gap in a critical data set is unacceptable. A strategy to replace a failed sensor is thus a critical component of any operational program. This replenishment strategy is also dependent on the definition of sensor failure, which in turn may be different for the research and operational user communities. For example, an imager might lose a few channels or the signal-to-noise ratio might increase, and yet the sensor might still return useful data from an operational perspective. However, such data could be useless for the more demanding needs of Earth system and climate researchers.

Data Continuity in Research

In Earth system research, the broadly ranging requirements for data continuity are based on an assessment of the critical scales of variability of the processes under study. For example, if ENSO forcing is important and ENSO events are assumed to occur generally every 3 to 5 years, then a gap of 0.5 to 1 year will likely not compromise the quality of the data set for climate research. On the other hand, a gap of 2 to 4 years will seriously degrade the quality of the record. Another example is the Antarctic ice sheet. It has been suggested that the ice sheet need only be mapped every few years and that a continuous record is not required. The risk in this thinking is that catastrophic events may be a critical component of ice sheet dynamics, and these events might be missed with such a sampling strategy. Continuity requirements, therefore, will depend on the science objectives and our understanding of the scales of variability.

Although the current EOS instruments were carefully designed, continuing improvements in technology can and will influence design trade-offs. There must be an effective mechanism for periodically revisiting these trade-offs and incorporating new technologies that can enhance capability or reduce cost. This introduces a classic cost trade-off: there are significant nonrecurring engineering costs associated with developing new instruments, so that the cost of developing the first of a series of new instruments is almost invariably higher than simply building another copy of an existing design. However, newer sensor designs could offer savings in total system cost if size, mass, and power reductions would permit corresponding reductions in satellite bus and launch vehicle size. An additional element in the trade is the potential nonrecurring investment to design the smaller satellite needed to realize these system-level savings.

In essence, there are competing strategies for effecting economies. On the one hand, the most powerful economic strategy is to produce many copies of the same design, thereby deriving economies of scale from quantity production and amortization of nonrecurring development costs over many units. On the other hand, advances in technology will eventually change the cost-versus-capability analysis enough to overwhelm economies of scale. An effective strategy must balance these countervailing trends by embracing a methodical approach that captures some economies of scale by producing several copies of a particular design, but then periodically introduces significant design alterations—block changes—that exploit advances in technology (perhaps on every second series of satellites). Indeed, this approach has been successfully used by other satellite programs that faced challenges similar to those of NPOESS.

The complex challenge of how to achieve technical renewal while maintaining data continuity and quality can best be addressed by embracing a concept the committee calls *dynamic continuity*. Specifically, the quality and continuity of measurements must be transparent to three levels of changes:

- Between successive flights of the same instrument design,
- Between successive generations of instruments, and
- Between similar instruments fielded by different countries.

Dynamic continuity can be achieved via a strategy that encompasses a rigorous calibration program, utilizing both in-orbit and in situ measurements, and a disciplined approach of overlapping measurements between successive generations of instruments over a sufficient interval to ensure accurate cross-calibration. In many respects, NASA did this very well in the Landsat (Land Remote Sensing Satellite) 4-5 era, where the then experimental thematic mapper instrument was carried alongside the existing operational multispectral scanner.

This same basic approach can and should be carried forward as a technology insertion strategy, although it is not strictly necessary to carry the old and the new simultaneously on the same spacecraft. Indeed, small satellites become an ideal vehicle for the development and introduction of new technology in a manner that does not introduce risk into the mainstream scientific measurement program. Specifically, new payload instruments, developed in parallel with EOS and flown on small satellites (perhaps in formation with the EOS spacecraft), would be an excellent approach for establishing the validity and comparability of new and old measurements. This approach also provides an opportunity to develop and prove the algorithms for reducing and analyzing the data from these new payloads and to provide an orderly transition to these experimental instruments, which are first flown on small satellites and then become the next-generation mainline instruments on subsequent spacecraft. In addition, it encourages the pursuit of high-risk, high-payoff technologies. By conducting the development off-line without risk to the operational mission, fear of failure does not become an overriding principle that stifles innovation. Moreover, the development protocol for such technology demonstrations can be less formal and expensive.

SIMULTANEITY

Some EOS and NPOESS requirements necessitate multiple, nearly simultaneous observations of the same location on Earth. In some cases, this requirement can be met with a single highly capable instrument. For example, the Moderate-Resolution Imaging Spectrometer (MODIS) will provide nearly simultaneous observations in the infrared and visible portions of the spectrum for studies of cloud properties and illustrates the simplest level of integration necessary to achieve simultaneity. More difficult challenges in achieving simultaneity arise when data streams from multiple sensors must be combined either to derive a geophysical quantity or to study specific Earth processes. These challenges can often be met by co-boresighting multiple sensors on a common satellite and—in many cases—by deploying the sensors on multiple platforms flying in formation.

Clouds are perhaps the most rapidly changing element of the Earth system: Cumulus cloud lifetimes can be as short as 15 minutes, and winds can move clouds at speeds greater than 1 km/minute. Therefore, measurements of the Earth's radiation budget (in which clouds play a dominant role) have the most stringent simultaneity requirements. The processes of cloud motion and cloud development will cause two different sensors to measure different portions of the cloud field. This discrepancy introduces random errors in the validation of cloud properties that might be derived from other sensors. More importantly, it introduces errors in the estimation of critical properties such as radiative fluxes within the cloud field. In the case of EOS platforms, this strict simultaneity requirement applies to measurements made by MODIS and CERES (Clouds and the Earth's Radiation Energy System) instruments. Researchers at NASA's Langley Research Center have performed analyses that account for two sources of co-location error: navigation errors between sensors and time differences between sensors. It was concluded that 6 minutes was the largest acceptable difference in time between the two sensors and that 3 minutes should be the goal.

Other sensors require contemporaneous observations but do not need to be exactly simultaneous; the interval between these observations can vary from minutes to days. For example, studies of linkages between wind forcing over the ocean and primary productivity require that measurements be made on the same day. The critical fact, especially for climate research, is that the measurements are made during the same time period and for a sufficient length of time to resolve the important scales of variability.

SAMPLING ERRORS

When observing any process, the quality of the measurement is a function of both the intrinsic characteristics of the sensor—such as its sensitivity and dynamic range—and its sampling characteristics. In remote sensing, the research community often emphasizes sensor performance, and technical innovation is thus driven to improve the sensor characteristics. Overall measurement quality is usually dominated by sampling errors, because the temporal and spatial scales of natural variability are not adequately resolved by the observing system. Although sampling theory will provide an initial quantitative estimate of the errors associated with a particular sampling strategy, evaluation of these errors depends in part on our understanding of the critical scales that must be resolved. For example, if we assume that ENSO events dominate primary production in the eastern tropical Pacific, we will design a sampling strategy to observe this system with acceptable error levels. But if smaller scale processes such as mesoscale eddies unexpectedly dominate the system, our observing strategy may not resolve them with sufficient accuracy.

Sampling requirements are fairly well understood in operational satellite systems. Long time records and an extensive history of numerical weather prediction have provided a foundation for the development of a robust observing system. Earth system processes, on the other hand, are not as well understood, especially in the area of feedbacks and linkages between the components of the Earth system. Therefore, an Earth observing strategy for research must balance quality of measurement with quality of sampling. Multiple copies of lower performance sensors on a constellation of small satellites may provide a better data product than a single high-quality sensor that cannot adequately resolve key processes. The appropriate strategy will depend on the processes and scientific questions under study; there is no one correct answer.

Increased sampling with multiple satellites does not necessarily improve the quality of the data product. The overall effect will depend on the details of the satellite sampling, the time and space variability of the geophysical field, and the scientific requirements. For example, Greenslade et al. (1997) analyzed satellite altimeter orbits in terms of their effective temporal and spatial resolution. Greenslade showed that the effective resolution is a function of the natural variability of ocean topography, the orbit characteristics of the platform (which drives sampling parameters such as repeat time), and the scientific criteria necessary to address a specific question (mean topography of the ocean). Although it would seem that multiple altimeters would automatically have better time/space resolution than a single altimeter, this was not the case. In fact, the orbits of the multiple platforms need to be studied carefully and coordinated in such a way that the sampling pattern does indeed improve the quality of the blended data product.

SUMMARY

Measurement requirements for operational programs and Earth system research reflect both scientific needs and the technical capability to acquire data of requisite accuracy and resolution. As scientific knowledge improves or as measurement capabilities increase, requirements can be expected to evolve. Although research and operational systems differ in their detailed requirements and approach, there are several common issues that must be addressed. These include instrument calibration, prelaunch sensor characterization, data record continuity, simultaneity, and sampling strategies. All of these issues influence mission architectures for small satellites, although the research and operational user communities may assign differing importance to them.

Requirements for spatial resolution, calibration, and other sensor performance criteria greatly influence sensor size. In turn, sensor size affects the overall size of the platform and the launch vehicle. More stringent requirements, which are often associated with research missions, lead to larger satellite solutions. In contrast, operational missions need to ensure continuity of critical measurements in a cost-constrained environment. These requirements often lead to a design based on multisensor platforms using a block purchase approach. Very different success criteria can thus lead to similar approaches to system architecture.

Small satellites can potentially alter significantly the approaches for both research and operational missions. Typically, the primary argument for small satellites in research is the ability to deploy low-cost missions, provide more flexibility in scientific and technical approaches, and obtain results more quickly because of a shorter development cycle. If small satellites and small launchers eventually follow a commodity pricing model (where the profit margin is small on individual units and revenue is generated from high-volume sales of these low-margin units), the science community will shift its perspective on remote sensing mission design.

Until now, most missions were designed to push the envelope in terms of technical ability and sensor performance. There was an implicit preference for a high-quality measurement once per day versus lower quality measurements several times per day. As the research community moves toward systematic observations of the Earth system, spatial and temporal sampling become more important factors. This is prompting a rethinking of the performance requirements of an individual measurement and the coverage requirements in time and space. Indeed, new satellite and sensor technologies are fostering a fundamental shift in Earth remote sensing measurement and satellite options. In particular, satellite constellations and clusters could provide significantly better coverage and open up new approaches for calibration and data continuity. The research community needs to evaluate the time and space scales of critical processes and match them with the appropriate sampling strategy. Such new satellite architectures no longer constrain the community to a single sampling approach, such as a Sun-synchronous orbit with a fixed equatorial crossing time.

Although this chapter's discussion takes essentially a research perspective, the time sampling strategy of operational missions such as NPOESS could also be analyzed rigorously. If the EDRs were prioritized, a strategy different from the present small constellation of three medium-sized platforms might result. **The committee recommends that both the research and operational communities perform a complete analysis of sampling strategies in the context of potential new mission architectures.** The result of this analysis might be a different mix of sampling strategies, including small one-of-a-kind missions, constellations of small satellites, and a few mid-size multisensor platforms. As discussed in subsequent chapters of this report, the maturation of remote sensing science and the development of new sensor, platform, and launcher technologies allow for a more systematic approach to both research and operational Earth remote sensing. Just as personal computers and networks have revolutionized the way computational systems are organized, new technologies in remote sensing can shift the way we design observing systems.

REFERENCES

Bruegge, C.J., D.J. Diner, and V.G. Duval. 1996. The MISR calibration program. J. Atmos. Oceanic Technol. 13(2):286–299.

Bruegge, C.J., V.G. Duval, N.L. Chrien, R.P. Korechoff, B.J. Gaitley, and E.B. Hochberg. 1998. MISR prelaunch instrument calibration and characterization results. IEEE Trans. Geosci. Remote Sensing 36:1186–1198.

Greenslade, D.J.M., D.B. Chelton, and M.G. Schlax. 1997. The midlatitude resolution capability of sea level fields constructed from single and multiple satellite altimeter datasets . J. Atmos. Oceanic Technol. 14:849–870.

Willson, R.C. 1997. Total solar irradiance trend during solar cycles 21 and 22. Science 277:1963–1965.

3

Payload Sensor Characteristics

The design of payload sensors (instruments) for Earth observation flows logically from the measurement requirements that have been established by the science and instrument teams. There is a straightforward process whereby the fundamental scientific questions to be answered lead to a set of measurement requirements that establish the point of departure for sensor design and trade-off studies. Trade-off studies assess instrument design alternatives that can produce the required measurements within the constraints set by fundamental physics, the state of the technology, and cost. Within these bounds, sensor designers work to minimize size, mass, and power requirements. Once established, these characteristics, in turn, determine the payload accommodation requirements. The sizing of the payload sensors(s) affects the design of a mission's entire space segment and establishes the scale (and therefore the cost) of the spacecraft and launch vehicle.

Chapters 4 and 5 of this report provide data showing that smaller satellite buses and smaller launch vehicles are available at lower cost (and capacity) than larger versions. It follows then that the smallest launch vehicle and smallest spacecraft bus that can accommodate a given mission sensor payload will generally yield the lowest mission space segment costs. Advances in sensor technology that can reduce instrument size, mass, and power consumption thus provide considerable leverage and are a frequent objective of sensor technology development efforts, which also aim at producing new or improved measurement capabilities and/or lowered costs. This chapter and Appendix B discuss the sensor design process and the potential for size, mass, and power reduction within the constraints of the fundamental physics that govern the desired measurements.

PAYLOAD DESIGN AND ACCOMMODATION REQUIREMENTS

The design process proceeds iteratively, as payload design engineers work with the scientific community to identify, challenge, and revise requirements that are driving cost, size, mass, and power. As the sensor design evolves, the requirements for accommodating the payload become apparent—that is, the mechanical, thermal, optical, and electrical interfaces that the spacecraft must provide in order for the payload to function as planned. Highlights of the process are illustrated in Figure 3.1.

The resulting interface requirements are then analyzed with the spacecraft engineers to arrive at a spacecraft and launch vehicle design that appropriately balances cost and risk. Box 3.1 categorizes interface, or payload accommodation, requirements.

PAYLOAD SENSOR CHARACTERISTICS

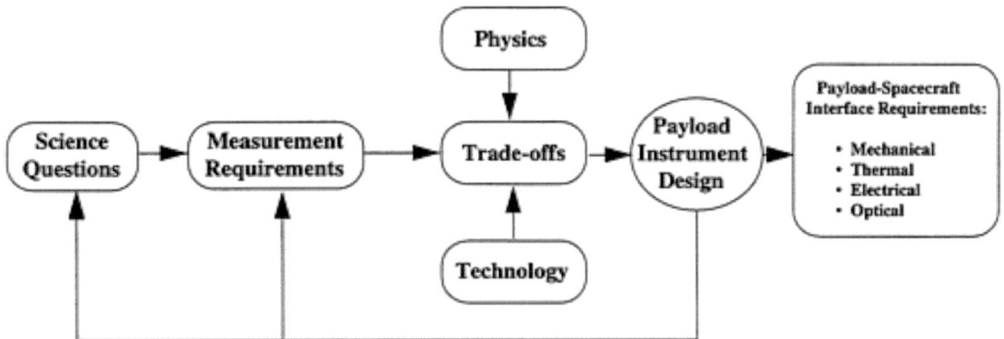

Figure 3.1
The payload design process generates requirements for accommodating the payload.

BOX 3.1 PAYLOAD ACCOMMODATION REQUIREMENTS

Mechanical

- Size (outline and mounting dimensions)
- Mass
- Moments of inertia
- Uncompensated momentum
- Launch loads (shock and vibration)
- Disturbances
  Thermal

- Conducted and radiated heat flux to/from payload
- Thermal gradients and baseplate distortion
  Electrical

- Power requirements
- Output data rate
- Command, control, and telemetry
- Electromagnetic interference
  Optical

- Sensor orientation and clear fields of view
- Pointing stability, agility
- Contamination: particulates, outgassing

Specific accommodation requirements vary widely depending on the type of payload instrument, with high-resolution, broad-swath, electro-optical multispectral imagers being the most demanding in terms of size, mass, data rate, and pointing accuracy and stability. Pointing and stability is often a differentiating factor between spacecraft designed to carry communication payloads and those designed to carry remote sensing payloads, the latter typically having more demanding requirements (see Chapter 4).

CURRENTLY PLANNED SENSORS

A comprehensive suite of payload instruments is currently under development for the National Aeronautics and Space Administration's (NASA's) Earth Science Enterprise (ESE)—including those that will be placed on the

Earth Observing System (EOS) AM, PM, and Chemistry satellites and the Tropical Rainfall Measuring Mission Observatory. TOMS (Total Ozone Mapping Spectrometer) and SeaWiFS (Sea-Viewing Wide Field-of-View Sensor) are other key sensors in the ESE. Efforts are also under way to define the payload instruments and develop the technology for the National Polar-orbiting Operational Environmental Satellite System (NPOESS), the next generation of polar-orbiting meteorological satellites. The spacecraft launches and corresponding payloads planned for the ESE are summarized in Figure 3.2.

The payload instruments represented in ESE and NPOESS programs cover a remarkably wide range of functions and interface requirements. These instruments include electro-optical and microwave radiometers, spectrometers, imagers, sounders, and altimeters that range from extremely compact to quite substantial. For example, some of the space environment monitoring instruments planned for NPOESS occupy less than 0.001 m^3 and have a mass of only 3 kg; in contrast, the three instruments composing the ASTER (Advanced Spaceborne Thermal Emission and Reflection Radiometer) and the MODIS (Moderate-Resolution Imaging Spectrometer), which are facility class instruments in the EOS program, have volumes and masses of 2.7 m^3/421 kg and 2.0 m^3/229 kg, respectively. Accommodation requirements for representative EOS and NPOESS sensors are tabulated in Table 3.1. In reviewing these data, it is evident that many payload instruments as currently designed could be flown on small satellites (100 to 500 kg); others are clearly unsuitable, even with advanced satellite buses that promise payload mass fractions of 40 percent or more.

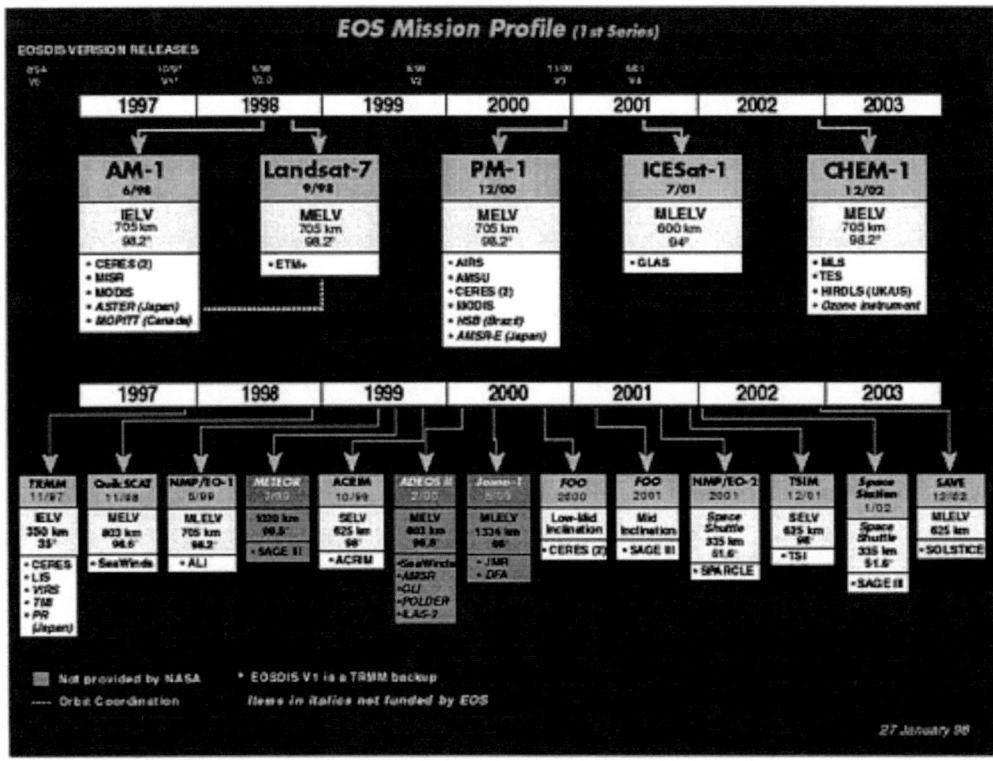

Figure 3.2
EOS mission profile (courtesy of NASA). Note that all acronyms are defined in Appendix E.

TABLE 3.1 Payload and Related Accommodation Requirements for Representative EOS and NPOESS Instruments

Instrument	Size[a] (mm) L or vel. dir.	W or Sun dir.	H or nadir dir.	Vol. (m^3)	Mass (kg)	Power (watts)	Data Rate (kbps) avg.	Notes
EOS INSTRUMENTS								
AIRS	1,397	775	762	0.83	156	256	1,440	
ASTER					421	463	8,300	
VNIR	580	650	830	0.31				
SWIR	720	1,340	910	0.88				
TIR	730	1,830	1,100	1.50				
CERES	600	600	576	0.21	50 per scanner	47 per scanner	10	
MISR	1,300	900	900	1.0	149	83	3,300	
MODIS	1,044	1,184	1638	2.02	229	163	6,200	
MOPITT	1,150	930	570	0.61	192	250	28	
NPOESS INSTRUMENTS								
IPO-Developed Instruments: Not-to-Exceed Values								
VIIRS[b]	1,320	1,290	1,380	2.35	200	300	8,000	0530 & 1330 Orbits
CrIS	640	450	470	0.135	81	91	1,500	1330 Orbit
CMIS[b]	63,500 max stowed dim.			TBD	275	340	500	0530 & 1330 Orbits
GPSOS[b]	330	230	270	0.02	22	40	50	0530 & 1330 Orbits
OMPS	450	540	560	0.136	45	45	50	1330 Orbit
SES	TBD	TBD	TBD	TBD	100	40	50	0530 & 1330 Orbits
Leveraged Instruments: Estimates for IPO Planning Purposes								
ATMS	700	600	400	0.168	66	85	300	1330 Orbit
TSIS	640	640	640	0.262	49	71	15	0530 Orbit only
ERBS	410	410	590	0.099	55	55	10	1330 Orbit only
ALT					98			Weight includes antenna;
Electronics	460	330	280	0.043		110	24	antenna dimension is a diameter.
Antenna	11,400	11,400	TBD					0530 Orbit
DCS					70	66	NA	
Electronics 1	200	360	280	0.020				
Electronics 2	200	310	280	0.017				DCS in 0530 & 1330 Orbits
Electronics 3	200	310	280	0.017				
Antenna	50	50	230	0.0005				L-W dimension is a diameter.
SARSAT					46	70	NA	0530 Orbit only
RPU electronics	390	280	200	0.022				
SPU electronics	310	280	200	0.017				
Antenna	80	80	650	0.003				L-W dimension is a diameter.

NOTE: All acronyms are defined in Appendix E. NA = not applicable; TBD = to be determined.

[a]These are Sun-synchronous satellites, Y-axis points in Sun direction.

[b]VIIRS, CMIS, and GPSOS are also planned for launch on a European METOP (meteorological operational) polar satellite. The METOP is a joint undertaking of the European Organisation for the Exploitation of Meteorological Satellites (EUMETSAT), the European Space Agency (ESA), and their member states.

SENSOR COSTS

Intuitively, it would seem that payload sensors should account for a large proportion of a mission's space segment costs, since the measurements they make represent the mission's purpose, and the other space segment elements (satellite bus, launch vehicle) are really in service to the payload. In reality, payload costs are typically between 30 and 70 percent of the space segment total, with the variation depending on the complexity of the sensors, the nature of the mission (research versus operational), and the development state of the mission elements. This last factor involves the need for nonrecurring development costs. Research instruments are frequently burdened with such costs in order to employ the most current technology to gain enhanced performance; they subsequently must often bear the cost of more extensive characterization, calibration, and validation efforts. Operational payloads incur nonrecurring development costs to achieve long-lived, high-reliability designs, but then typically benefit from block buys where only the costs of manufacture, integration, and testing are incurred for subsequent units.

Operational sensors are often derived from successful research sensors and benefit from their already-paid-for development costs. Similarly, research sensors often benefit from extensive, separately funded technology development that precedes the actual sensor development program and is not booked against it. True sensor costs are hard to ascertain even for specific sensor programs. Sensors under development can only show estimates of costs. For completed sensors, costs are sometimes allocated to multiple accounts, which can further obscure actual costs.

The committee encountered these and other difficulties in trying to understand and compare the costs for EOS research sensors with their NPOESS counterparts. The most direct comparisons were possible for the programs' multispectral imagers (MODIS and VIIRS [Visible and Infrared Imaging Sensor]) and infrared sounders (AIRS [Atmospheric Infrared Sounder] and CrIS [Cross-track Infrared Sounder]), as shown in Table 3.2.

The data show that the target first unit costs for the operational NPOESS sensors are substantially lower than the actual costs of their EOS counterparts. One reason for this cost differential is that the first unit costs for the NPOESS sensors include those for the data processing algorithms, whereas the costs for the EOS sensors do not. The data also show that the recurring target costs for subsequent NPOESS sensors are much lower than for their EOS counterparts.[1] Part of the anticipated reduced costs for the NPOESS sensors reflects (1) diminished requirements as a result of reduced capability (e.g., VIIRS has less on-board calibration capability than MODIS, and CrIS has fewer spectral channels than AIRS); (2) a more efficient, multiple-buy procurement process; and (3) prior development (of the EOS and other sensors) and advances in technology and processes. With proper funding to enable them, advances in sensor technology and processes should continue to improve performance, cost, size, mass, and power parameters to suit mission objectives.

FUTURE SENSOR DESIGNS: IMPLICATIONS OF ADVANCED TECHNOLOGIES

Size and Design Constraints[2]

The "trade space" for the design and sizing of payload instruments is established by four types of constraints: fundamental, technological, mission, and programmatic (see Box 3.2). As technology development progresses in critical payload components and materials, the degrees of freedom available to the payload designer increase so that more efficient and capable instrument designs become feasible. There are limitations, however, because not all elements of a payload are amenable to miniaturization. While we can expect continued improvements in the size/mass/power efficiency of payload electronics, for example, the size of other subsystems—notably optics and radiative coolers—are set by first-order physics that are not subject to technological improvement. Even for subsystems and components that are driven by physics, improvements in materials and packaging techniques can

---

[1] Although there are no plans to build a second AIRS unit given the cost of the first unit, it is likely that a second would cost considerably more than the $12.5 million recurring cost target for CrIS.

[2] The information in this section provides an overview of the physical limits on sensor sizing. Appendix C continues this discussion and examines in greater detail the prospects for building more compact sensors for EOS and NPOESS.

significantly affect size/mass trade-offs. For example, relatively large apertures are feasible with small satellites by invoking deployable or inflatable technologies, although there are attendant costs and risks.[3]

Table 3.2 EOS and NPOESS Sensor Cost Comparisons

	First Unit Costs ($M)	Subsequent Unit Costs ($M)
MODIS (EOS)	183.8	90.8
VIIRS (NPOESS)	150.1	27.5
AIRS (EOS)	247.0	None planned
CrIS (NPOESS)	67.7	12.5

NOTE: EOS data are actual costs; NPOESS data are target costs.
SOURCES: EOS data—Michael King, NASA Goddard Space Flight Center; NPOESS sensor cost data—Stan Schneider, National ceanic and Atmospheric Administration NPOESS Integrated Program Office.

BOX 3.2 FOUR TYPES OF PAYLOAD DESIGN CONSTRAINTS

Fundamental	Diffraction limit, photon noise, Nyquist limit, radiative-transfer limit (Planck's law) . . .
Technological	Detector size and performance; optical form, figure, fabrication, and alignment; processor speed . . .
Mission	Size, weight, power constraints due to spacecraft/launch vehicle; link capacity . . .
Programmatic	Cost, schedule, and risk constraints; legislative and regulatory requirements

---

Within the constraints outlined above, sensor design and size are determined by a complex trade-off among spatial, spectral, and radiometric performance. All three of these performance measurements are interdependent, and, barring compensatory design changes, each performance measure is improved at the expense of the others. Developing a design that balances all of these performance parameters while minimizing size, cost, and technical risk is the essence of sensor system engineering.

Fundamental Limits on Size

The physics of diffraction and radiometry's energy collection requirements establish the fundamental limits on aperture size. Aperture size, in turn, is often the single most important parameter in determining the scale of a remote sensing instrument. Such a general pronouncement has many exceptions, however; there are sensor designs, for example, in which the swept volume of a scanning mirror or the size of a passive radiative cooler becomes the dominant factor.

Technological Limits on Size

The theoretical performance limit on radiometric sensitivity provides a useful framework for examining the interdependencies among performance and design parameters. Radiometric sensitivity (signal-to-noise ratio) depends on the optical aperture area, the detector solid angle (spatial resolution), signal integration time, and spectral bandwidth, among other parameters, as noted in Figure 3.3.

---

[3] Both NASA and the Department of Defense are currently developing large aperture deployable and inflatable technologies, ranging across the spectrum from radio frequency (e.g., Very Long Baseline Interferometry) through visible (e.g., Next Generation Space Telescope) wavelengths.

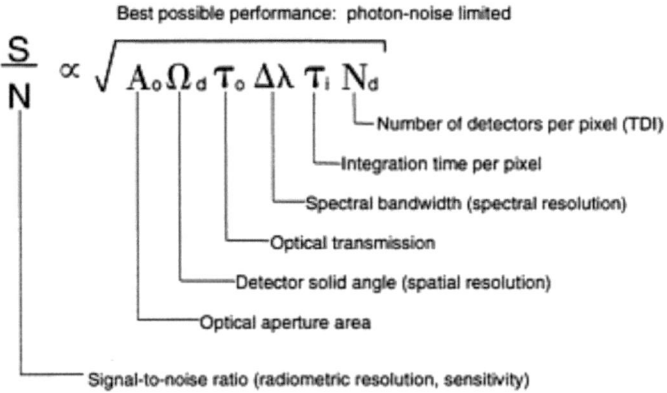

Figure 3.3
Factors determining detector signal-to-noise ratio.

Improvements in either spatial or spectral resolution thus come at the expense of radiometric sensitivity unless these are offset by a larger aperture, better optical transmission, or increased integration time. Improvements in technology can influence the relative attractiveness of these alternatives. For example, effective integration time (and hence radiometric sensitivity) can be increased by adding more detectors and slowing the effective scan rate or by using additional detectors to take multiple samples of the same point in object space and coherently summing the outputs—a process called time-delay integration (TDI). These strategies offer improved performance at the expense of added complexity in the detector arrays and associated signal processing electronics; on the other hand, technology improvements in these areas have made it possible to exploit such design strategies with relatively low cost and risk.

The foregoing discussion offers a good example of the way in which technology advances have led to changes in the design optimization process. In the 1970s, detector technology and signal processing electronics were less well developed, so intelligent designs of that era minimized the number of detectors via the use of scanning mechanisms and relatively large optical apertures. Now that higher density detector arrays and sophisticated signal processing electronics are readily available, aperture size can be (and has been) significantly reduced for many classes of instruments. Still, for imaging sensors of relatively high spatial resolution, aperture diameter is clearly the dominant sizing parameter because diffraction-limited performance is a fundamental physical limit that is not amenable to technological improvement. Technology advances have nonetheless made it feasible to design systems whose performance is close to theoretical limits.

Figure 3.4 shows the relationship between aperture diameter and spatial resolution (ground sampling distance) for different spectral wavelengths and altitudes.[4] For example, an aperture diameter on the order of 10 cm is needed to provide 10 m spatial resolution from an altitude of 705 km. Curves such as these can be misleading, however, because "resolution" is not meaningful unless defined in terms of an equivalent modulation transfer function or point spread function. Indeed, there could be a four-to-one difference in aperture diameter for different sensors all purported to have the same spatial resolution, as denoted by the extent of the design range bar in Figure 3.4. Sensors at different points within that design range would exhibit substantial differences in performance. These differences arise from adopting different design approaches in balancing sensor subsystem performance.

---

[4] Ground sampling distance is the center-to-center distance between successive pixels imaged on the ground and can be expressed either linearly (typically in meters) or angularly (typically in microradians)—the angular measure is invariant with altitude, while the linear measure is altitude dependent. The term is used interchangeably with "ground sampling interval," although the latter is less commonly used.

PAYLOAD SENSOR CHARACTERISTICS

Sensor designers can choose the governing factor that establishes the spatial resolution of an instrument: either the optics or the detector can become the limiting element. Three different design philosophies are illustrated in Figure 3.5. All three systems might claim to have 10 m spatial resolution, for example, because each of these designs uses a detector element that has a 10×10 m projection on the surface of Earth.

In the first case, performance is truly limited by the optical aperture: The diffraction-limited blur circle of the optics is the largest (dominant) component. This design point represents the lower extreme of aperture sizes in the design range shown in Figure 3.4. Such an approach extracts the most performance from a given aperture size. Because the detector size is much smaller than the optical blur circle, the resulting point spread function (the convolution of the optical blur and the detector's active area) is nearly the same as the optical blur alone. Note that in this case, however, it would be highly misleading to claim that the resulting sensor had a spatial resolution equal to the size of the detector. Rather, the equivalent size of a resolution element would be more like the size of the optical blur circle—about one-half the resolution claimed.

The second example in Figure 3.5 shows a design case that balances the optical blur circle and detector size so that they are approximately equal. This is the design point that corresponds to the center of the design range in Figure 3.4. The third example, with the smallest optical blur circle, corresponds to the largest aperture diameter within the design range. This design would have the best spatial resolution of the three alternatives, but it would also be the largest of the sensor designs.

Measurement Strategies and Mission Architectures

Payload sizing and the feasibility of employing small satellites depend as much on measurement strategy and architecture as on specific sensor technologies. Different approaches to partitioning measurements among different instruments and satellites can have significant effects on the size/mass/power of payload sensor designs and their corresponding host spacecraft. This becomes the classic trade-off between fewer, highly capable multipurpose instruments versus a greater number of smaller, simpler, more specialized instruments tailored for specific measurements.

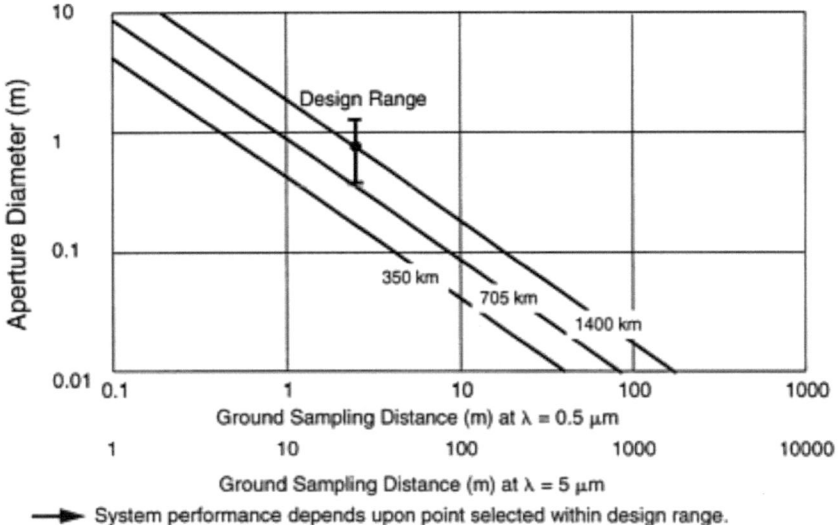

Figure 3.4
Diffraction dictates aperture size.

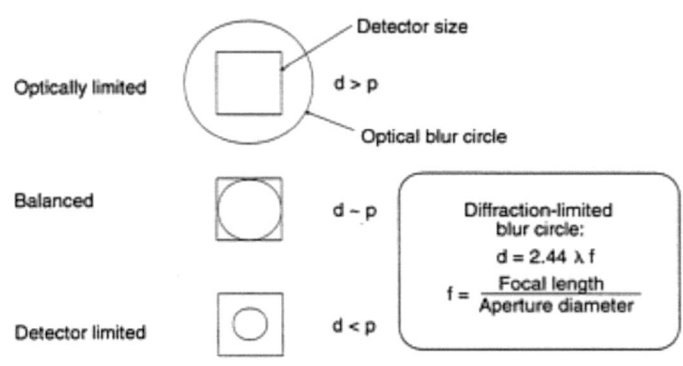

Figure 3.5
Different design approaches yield a 4:1 range in aperture size (or equivalently in focal ratio).

For example, MODIS, with its 36 spectral channels ranging from 0.405 μm to 14.385 μm, makes a variety of measurements that serve a broad array of applications including vegetation/biomass estimation, cloud cover, atmospheric sounding, and ocean color. These measurements could be acquired by two, three, or perhaps four separate, smaller sensors tailored for a specific application. Each of these smaller instruments could even fly on its own dedicated small satellite (or even smaller microsatellite), though the economic advantage of designing and building a proliferation of smaller sensors is not clear. Small satellite technology is at least an enabler that makes it feasible to address these trade-offs in search of a robust and economical solution to mission-level measurement needs.

SUMMARY

Sensor design should flow from measurement requirements and is best performed as an iterative process between designers and the science community within the constraints set by fundamental physics, the state of the technology, and cost.

The size of payload sensors affects the design of the entire space segment and establishes the scale of the spacecraft and launch vehicle. Through this relationship, the payload sensors exert substantial leverage over all space segment element costs.

It is often difficult to determine the actual costs of developing new sensors. Many development programs build on prior technology or sensor development efforts but do not account for these costs. Nevertheless, the National Oceanic and Atmospheric Administration anticipates lower costs for the NPOESS sensors compared with their heritage sensors on EOS. Lower costs might be expected due to the EOS development heritage, further advances in sensor technology, and more efficient procurements (e.g., multiple buys).

As currently designed, many—but not all—of the sensors planned for EOS and NPOESS satellites could be physically accommodated on small satellites in the 100 to 500 kg class. Chapter 6 provides a detailed analysis of the factors to be considered in determining whether it is cost-effective to distribute planned sensors on a larger number of satellite platforms. Larger, multipurpose sensors can be subjected to the classical analysis of the trade-offs in partitioning required measurements among several smaller sensors that could then be accommodated on several small satellites, although the economic advantage of doing so is not apparent.[5]

---

[5] Similar architectural trade-offs are discussed in Chapter 6 at the spacecraft level.

4

Small Satellite Buses

CAPABILITIES OF SMALL SATELLITE BUSES

Satellites are frequently described in terms of a payload and a service module or "bus." The capability of a satellite bus relates to its ability to accommodate payloads and to meet mission requirements. Payload accommodation requirements are many and include mass; geometry (volume, mechanical interfaces, fields of view); thermal interfaces; power (wattage, voltages, duty cycles); data (rates, interfaces); contamination environment; electromagnetic interference limits; and spacecraft pointing knowledge and control (slewing and settling rates, stability, jitter).

The mission architecture places further requirements on the spacecraft bus such as on-board data processing; data memory and communication links; battery capacity; and the need for propulsion (orbit insertion, orbit maintenance, formation flying, end-of-mission de-orbit). Additional mission requirements include spacecraft life (expendables, radiation dose, solar array degradation); reliability; and degree of redundancy.

All space missions are constrained by launch vehicle performance (mass to orbit) and fairing—i.e., the aerodynamic cover that protects the spacecraft as it travels through the atmosphere—volume. These constraints can be severe for small expendable launch vehicles such as the Pegasus (see Chapter 5) and can lead to complex designs for "deployables" (such as the solar panels) in order to stow the satellite within the fairing, as in the Air Force Space Test Experiment Program Mission 1. Within these constraints, the satellite designer generally wants to maximize resources available to the payload and minimize those required for the spacecraft bus. Consequently, much small satellite technology development effort has been directed toward reducing bus volume, mass, and power consumption, while providing robust capability by increasing battery capacity, solar array efficiency, data memory, processing rates, and so on (NRC, 1994). This trend is likely to continue in avionics as well as in the still embryonic field of microminiature electromechanical systems.

Partly because of substantial investments by the National Aeronautics and Space Administration (NASA), Department of Defense, Department of Energy, and industry over the past decade, small satellite technology has already advanced to the point where a great deal of capability can be provided in a relatively small package. Table 4.1 shows typical performance parameters for current low- and high-end small spacecraft buses derived from information presented in NASA's Rapid Spacecraft Acquisition contract offerings. These are all off-the-shelf flight-qualified spacecraft buses NASA is making available to potential users through the Goddard Space Flight Center's Rapid Spacecraft Development Office (RSDO). Most are available in a basic or "core" version with options for enhancing performance.

Table 4.1 Characteristics of Small Satellites

Parameter	Low-End Buses (w/o options)	High-End Buses (w/ options)
Design life (years)	1–3	>>5
Reliability (at design life)	.8–.9	.8–.9
Avionics redundancy	Limited	Extensive to full
Bus mass (kg)	150–300	425–650
Payload mass (kg)	100–300	300–500
Payload power (orbital average, W)	60–125	100–500
Propulsion authority (kg Hydrazine)	0–25	33–75
Pointing accuracy (deg 3-sig)	0.02[a]–.25	0.01[a]–0.03[a]
Pointing knowledge (deg 3-sig)	0.001[a]–0.2	0.003[a]–0.008[a]
Data storage (Gbit)	2–64	12–200
Downlink (Mbps)	2–4 at S-band; 100 at X-band available on SA200S	2 at S-band, 320 at X-band

NOTE: The low-end buses are the Spectrum Astro SA200S, Swales, and the three-axis TRW STEP; the high-end buses are the Ball RS2000, Lockheed Martin LM900, and TRW SSTI-500.
[a] With star trackers.
SOURCE: RSDO (1999).

This level of performance, especially at the upper end, is adequate to support many—but not all—Earth observation missions. Some payloads are simply too large, too heavy, too demanding of power, or have moving parts that create too large a vibration source to be accommodated efficiently with a small satellite on a small launch vehicle (e.g., the Multifrequency Imaging Microwave Radiometer, Atmospheric Infrared Sounder, and Microwave Limb Sounder). Excessive payload size and weight can be addressed to some extent with more capable launch vehicles (e.g., NASA's FUSE [Far Ultraviolet Spectroscopic Explorer] mission), and needs for greater payload power with larger, more efficient solar arrays and higher capacity batteries. However, the limited inertia of a small satellite makes it difficult to control jitter without active isolation of large vibration sources.

Table 4.1 shows that small satellites can provide robust capability with respect to data storage and downlink rates. However, a proliferation of small satellites in orbit will raise ground station capacity and frequency allocation issues. High-data-rate ground receiving stations are limited in number, and new ones are costly to install and support. Competition for frequency allocation is increasing around the world; this process is limited and controlled by the Federal Communications Commission for the United States and by the World Administrative Radio Conference internationally. With the number of satellites increasing, the competition for ground station contact time and uplink/downlink frequencies and the potential for interference are also increasing. These problems are an important aspect of the trade-offs entailed in system design and mission planning.

SPACECRAFT BUS COSTS

The cost of small spacecraft buses is a somewhat elusive parameter, depending as it does on the capability of the bus, the technological heritage, and the details of program management and bus production. Currently, recurring costs for spacecraft buses like those in Figure 4.1 range from approximately $10 million to $30 million. Nonrecurring costs can add another 150 percent if a complex, new, mission-unique bus must be developed, but substantially less if previously developed spacecraft can be adapted for use.

A recent RAND study on small spacecraft offers some interesting perspectives on spacecraft costs (Sarsfield, 1997). Traditionally, cost modelers have used a cost estimating relationship based on mass to predict spacecraft development costs. However, as shown in Figure 4.1, variation in development costs for small spacecraft is much greater than for larger spacecraft. Very low costs—a key objective in the push toward small satellites—are experienced only with very simple spacecraft performing very limited missions. Small spacecraft can be relatively expensive when they retain the complexity required to meet demanding scientific objectives (pointing accuracy, power, processor speed, etc.). For demanding missions, development costs are a relatively weak function of spacecraft mass; thus, specific cost (cost per unit mass) increases as mass decreases.

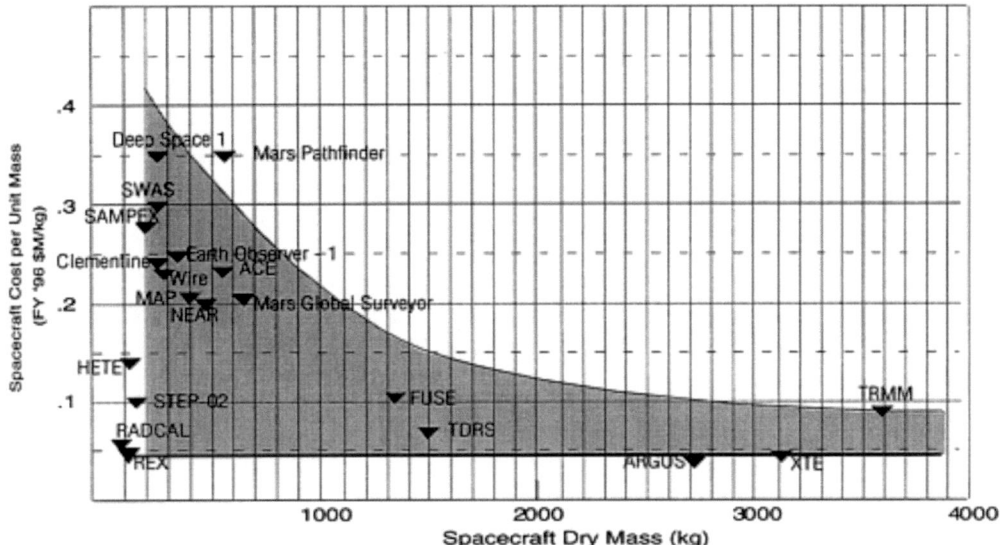

Figure 4.1
The relative cost of small spacecraft (adapted from Sarsfield, 1997). Note that all acronyms are defined in Appendix E.

The RAND study also addresses the impact of development processes on spacecraft costs (Sarsfield, 1997). Streamlining the development process offers cost efficiencies for spacecraft of all sizes. However, it is easier to streamline processes with smaller, simpler satellites involving smaller development teams. Process-driven spacecraft cost reductions achieved with small satellites have ranged from 15 to 30 percent. Programmatic issues associated with small spacecraft missions are further discussed in Chapter 7.

UTILITY OF "COMMERCIAL" SPACECRAFT

Recently, there has been great interest in the possible use of commercial spacecraft buses to perform science missions as a way of avoiding or reducing bus development costs. As used here, "commercial" spacecraft buses are those for which there exists an operating production line serving a commercial market, as is the case for several manufacturers of communication satellites (e.g., Iridium). It is important to differentiate "commercial" from "standard" buses. Several suppliers of small satellites offer a standard bus consisting of flight-qualified components configured for the particular mission at hand. These offerings generally involve a core bus plus a range of options to increase (or decrease) capability. As such, they really represent standard architectures employing standard spacecraft subsystems with defined interfaces. As opposed to a production line, such spacecraft are typically developed as individual projects by small teams co-located for efficiency. Several of the bus offerings available through NASA's RSDO (Table 4.1) fall into this category.

The standard bus approach goes a long way toward lowering costs by reducing—but not eliminating—nonrecurring development. The commercial bus offers the potential for even greater cost reduction; if directly applicable to the mission, almost all nonrecurring development costs are avoided, and the recurring costs of manufacture benefit from the efficiencies of the existing production line.

For defined payloads, most missions will not be able to use a bus directly off the line. Rather, in the great majority of cases, the bus will have to be tailored to the mission with some degree of modification. For example, Lockheed Martin modifies its LM700 Iridium bus for more demanding scientific missions (LM700B) and offers others (LM100 and LM900) for Earth observation missions in the NASA RSDO catalog.[1]

Recently, NASA sponsored a study of alternative architectures for performing the Earth Observing System Chemistry-1 mission. Eight suppliers with existing spacecraft buses performed studies to accommodate the HIRDLS and ODUS (High Resolution Dynamics Limb Sounder and Ozone Dynamics Ultraviolet Spectrometer), TES (Tropospheric Emission Spectrometer), and MLS (Microwave Limb Sounder) instruments on three spacecraft. No supplier had an existing commercial spacecraft bus that could perform these missions without significant modification and attendant costs. The question remaining then was the level of nonrecurring costs needed to modify a commercial bus versus those to configure a standard bus to meet mission requirements. The answer is mission specific and will be determined by the marketplace. In the case of the Chemistry-1 studies, cost estimates for the two cases were similar.

SPACECRAFT CAPABILITY AS A PAYLOAD DESIGN PARAMETER

An alternative paradigm that has been suggested is to transfer the burden of accommodation from the spacecraft to the payload; that is, to treat the spacecraft capabilities as payload design requirements much as the spacecraft designer currently treats launch vehicle capabilities as design requirements. This approach minimizes the costs of using either commercial or standard spacecraft buses. It also introduces several vexing issues:

- It places an increased burden on the payload developer who is often less experienced in space systems than are spacecraft manufacturers.
- Instrument development cycles (4 to 5 years) are typically much longer than those for small spacecraft (18 to 36 months). Thus, to define requirements, payload developers must select candidate spacecraft buses early in their design cycle and must somehow ensure availability when needed.
- More than one sensor supplier makes it very difficult to configure missions. System design integration for multisensor payloads is traditionally performed by the satellite manufacturer, which must ensure that all sensors are accommodated without interference.

PRINCIPAL INVESTIGATOR-LED PROJECTS

Recent NASA procurements (Discovery, Mid-size Explorer, Earth System Science Pathfinder, and Small Explorer) have embraced a "PI mode" wherein a principal investigator (PI) configures and leads a team to propose and compete for missions in response to a fairly broad Announcement of Opportunity (AO). In many cases, interested PIs have solicited industrial teammates with existing spacecraft buses to join their teams, thus defining the spacecraft capabilities and payload accommodation requirements early enough in the process to achieve efficient interface compatibility.

This approach works when the PI has a sufficiently well-defined payload to select an appropriate bus (and hence teammate) at the time of the procurement. Most proposals submitted in response to recent PI mode AOs have involved payloads that have been under development for some time and are relatively mature. For example, the fourth Discovery mission—Stardust—carries aerogel capture cells proven on numerous Get Away Special Sample Return Experiments, and a camera that uses spare parts from the Voyager and Galileo missions. It remains

---

[1] The catalog can be found online at <http://rsdo.gsfc.nasa.gov/rapidi/catalog.cfm>.

to be seen how well the PI mode will work once the current inventory of sensor concepts, designs, and—in some cases—hardware have been exhausted.

FUTURE TRENDS

A key factor in the movement toward smaller spacecraft is the desire to reduce total mission costs.[2] Smaller satellites can cost less, particularly if the mission payloads have less demanding accommodation requirements. Also, smaller satellites can be launched on smaller launch vehicles which, setting current issues of reliability aside, offer a lower cost to orbit. Cost reduction will continue to be a driver for small satellite missions and the spacecraft buses that support them. Continued technological development will further increase capability and lower costs. NASA's New Millennium program is an important vehicle through which new technologies will be validated through flight demonstrations.

Avenues for additional cost reduction strategies (applicable to both large and small satellites) include the following:

- Streamlined procurement practices (PI mode, streamlined contractor practices, etc.);
- Low-overhead management techniques (concurrent engineering, integrated product development teams, customer insight through participation in lieu of oversight, reduction in formal reviews);
- Design and development improvements (computer-aided design, early definition of design requirements and interface control documents, hardware and software reuse, selected redundancy, spacecraft standards, and commonality); and
- Lowering the cost of operations (spacecraft autonomy, on-board processing).

Several of these strategies are discussed further in Chapter 7.

An excellent example of the trend toward streamlined procurement practices was NASA's Rapid Spacecraft Procurement Initiative (NASA, 1997). Through this solicitation, Goddard Space Flight Center has developed a "catalog" of industrial commercial and standard spacecraft bus offerings that can be quickly procured through indefinite delivery, indefinite quantity (IDIQ) contracts with seven suppliers. When applicable to the mission, these IDIQ contracts and catalog provide an efficient way for PIs to select industry partners when responding to AOs and for flight projects to acquire spacecraft buses. The goal is to shorten the procurement cycle from 9 to 12 months down to 30 to 120 days. This procurement approach was implemented for two missions—QuikSCAT (Quick Scatterometer) and ICESat (Ice, Cloud, and Land Elevation Satellite)—as of June 1998, with four more in competition and six (three outside NASA) under consideration (RSDO, 1999). On the other hand, an attempt to use this approach on SOLSTICE/SAVE (Solar-Stellar Irradiance Comparison Experiment/Solar Atmospheric Variability Explorer)—a program already well under way—was not successful. None of the standard bus offerings could accommodate the payload without employing a larger than desired launch vehicle with an unacceptable increase in cost.[3]

SUMMARY

Small satellite technology has advanced to the point where very capable buses are currently available for performing many Earth observation missions. However, some Earth observation payloads are too large, too heavy, too demanding of power, or generate too much vibration to be efficiently accommodated with small satellite missions.

Very low costs—a key objective in the push toward smaller satellites—are experienced only with simple spacecraft performing limited missions. Small spacecraft can be relatively expensive when they retain the complexity required to meet demanding science objectives (pointing accuracy, power, processor speed, etc.).

---

[2] An October 1996 National Research Council workshop examined ways of reducing the cost of science research missions (NRC, 1997).

[3] Daniel Baker, University of Colorado, Boulder, personal communication.

Commercial production buses offer the potential for reducing costs. However, they generally have to be tailored—with attendant costs—to accommodate existing Earth observation payloads. Designing new payloads to match existing bus capabilities offers greater cost-effectiveness, but caution must be exercised not to compromise the scientific mission in doing so. NASA's Rapid Spacecraft Acquisition Initiative exemplifies an innovative approach to matching existing spacecraft buses to payload accommodation requirements.

Efforts should continue to reduce bus volume, mass, and power, and to increase communications and data handling capacity, such that larger fractions of launch vehicle performance and fairings are made available to more demanding payloads. Lowest cost will be achieved when satellite size is matched to payload requirements and launch vehicles are matched to the satellite.

REFERENCES

National Aeronautics and Space Administration (NASA). 1997. Rapid Spacecraft Procurement Initiative, RFP 5-02816-001. Greenbelt, Md.: NASA Goddard Space Flight Center.
National Research Council (NRC). 1994. Technology for Small Spacecraft. Washington, D.C.: National Academy Press.
———. 1997. Reducing the Costs of Space Science Research Missions. Washington, D.C.: National Academy Press.
Rapid Spacecraft Development Office (RSDO). 1999. Available online at <http://rsdo.gsfc.nasa.gov>.
Sarsfield, L. 1997. The Cosmos on a Shoe String: Small Spacecraft for Space and Earth Science. MR-864-OSTP. Santa Monica, Calif.: RAND, Critical Technologies Institute.

5

Small Launch Vehicles

The effective use of small satellites to fulfill Earth observation needs depends on the availability and costs of launch vehicles. As spacecraft become smaller and less expensive, so also must launch vehicles, or launch costs will become disproportionately large. Additionally, the trend toward smaller spacecraft implies a commensurate increase in the number and rate of launches. It is well known that long launch queues, slips, and delays can increase overall mission costs. Similar problems with small missions are likely to have even greater impacts due to limits on launch site capacity.

Selecting the appropriate launch vehicle for a particular mission involves mission architecture trade-offs reflecting the number of satellites to be launched, orbit requirements, satellite on-board propulsion, and launch vehicle performance. Missions that call for multiple satellites in common orbit planes can accrue cost benefits with multisatellite launches on higher performing launch vehicles. Satellites that must carry on-board propulsion for orbit maintenance or attitude control can sometimes effectively exploit lower performance, lower cost launch vehicles to place them into low initial orbits and then use their own propulsion systems for final orbit insertion. Whatever the specifics, the launch vehicle must be matched to the mission if costs are to be minimized. Excess launch capacity beyond prudent margins represents wasted costs.

Recognizing that the move toward smaller spacecraft places added emphasis on the costs and availability of appropriate launchers, the aerospace industry has moved to develop a number of "small" launch vehicles tailored specifically to meet this growing market segment. This chapter presents an overview of these small launchers in terms of their known costs, performance parameters, capabilities, and performance records.

U.S. launchers were emphasized in this assessment, since current U.S. policy precludes the launching of government-funded spacecraft on foreign launch vehicles. Launchers based on converted ballistic missiles were also excluded on the grounds of current U.S. policy. Formal policy states that the use of converted ballistic missiles is restricted to government payloads only, and then only when such use would result in significant savings over the use of commercial launch services. Statements by administration personnel indicate that any requests to use a converted ICBM (Intercontinental Ballistic Missile) for an orbital flight would meet with strict scrutiny. In general, the National Space Transportation Policy directs U.S. departments and agencies to purchase commercial launch services to the fullest extent feasible. In the event that U.S. policy changes, some discussion of foreign launch vehicles is provided for a more complete assessment of small satellite launch capabilities.[1]

---

[1] The situation with launchers is changing rapidly. The military, among others, is petitioning to allow use of foreign launchers; at the

SMALL LAUNCH VEHICLES FOR EOS AND NPOESS

This section covers launch vehicles capable of launching to the Earth Observing System (EOS) and National Polar-orbiting Operational Environmental Satellite System orbits with mass performance capabilities up to and including the Delta II. While the Delta II may be considered excessive for the launch of individual 500 kg payloads (the upper limit of what this report defines as a small satellite), its capacity for launching multiple small spacecraft on a single launch vehicle merits its inclusion. Also, Boeing Corporation (which recently acquired McDonnell Douglas Co.) is developing a downsized version of the Delta II (Delta II-7320) to extend the utility of this reliable launch vehicle. However, this will still be a fairly high-performance launch vehicle with a relatively high absolute cost compared with the alternatives, suitable primarily for medium-sized or multiple small satellites. Within these guidelines, the launch vehicles considered here are the Delta II, Pegasus, Taurus, Athena (previously known as the Lockheed-Martin Launch Vehicle), and Conestoga. Further detail on these launch vehicles is provided in Appendix C, which also addresses the Eagle family of launch vehicles—the Eclipse Express and Astroliner, the PacAstro, and the Kistler booster. These launch vehicles, while all still in development, are included because of their potential for significant cost savings and market impact.

Pertinent data for the U.S. launch vehicles evaluated are presented in Tables 5.1 and 5.2. Table 5.1 summarizes their mass performance to a 700 km polar Sun-synchronous orbit, approximate cost, and performance history; Table 5.2 provides data on their fairing dimensions and launch environments. Table 5.1 also provides data for relevant foreign launch vehicles.

Generally, mission planners look to minimize mission costs. Because absolute launch vehicle costs increase with launch vehicle size and performance, the lowest performance (and hence lowest cost) launch vehicle that accomplishes the mission should be used. Preferably, the mission designer would have a series of launch vehicle options with increments in performance filling the gap between the low-capacity Pegasus and the high-capacity Delta II. Small launch vehicles such as the Pegasus and Athena 1 have limited capacity to put payloads into EOS orbit. However, these launch vehicles can be used for Earth observation missions by supplementing them with spacecraft on-board propulsion to enable them to reach the desired orbit (e.g., the Total Ozone Mapping Spectrometer Earth Probe). This approach is being used, but it results in some increase in spacecraft cost. The development of intermediate-capacity launch vehicles, such as the Taurus XL and Athena 2, helps fill this gap and offers more opportunity to optimize missions.

Fairing size is sometimes another criteria in selecting a suitable launch vehicle for a mission in that it must accommodate the stowed payload. It is preferable that launch vehicle candidacy not be limited by fairing size but by performance to orbit. Thus, most manufacturers are developing larger fairings for their vehicles for added utility. The fairing size for the Pegasus, however, which does impose significant size constraints, is limited by its airplane launcher system.

Figure 5.1 plots the cost and performance data for operational and planned U.S. launch vehicles as the specific cost per unit payload (satellite) mass to the EOS orbit versus launch capacity. For operational launchers, the minimal cost per unit mass to orbit is achieved with the Delta II and increases with decreasing or increasing launch vehicle capacity. The cost per pound penalty is severe for small launchers with payloads under 500 kg. It is this superior cost efficiency of the Delta II, along with its excellent reliability, which makes launching multiple satellites on a single Delta II an attractive alternative to multiple smaller launch vehicles when possible. In fact, early experiences (failures) with new, smaller launch vehicles indicate that reliability is a major concern, as indicated by the success rates shown in Table 5.1. It will probably take several years and more failures before any small launch vehicle achieves the reliability of the Delta II (>95 percent).

---

same time, a major issue is cooperation with and technology transfer to China. Complicating this issue is a policy debate within the administration and the Congress on how to balance competing economic and foreign policy interests with concerns over technology transfer—issues that resonate particularly with respect to China.

Table 5.1 Launch Capacity to EOS Orbit, Cost, and Performance History of Candidate Small Satellite Launch Vehicles

Vehicle/Configuration	Capacity to 700 km Sun-Synchronous Orbit (kg)	Cost ($M)	Performance History (Successes/Flights through Oct. 1988)
U.S. LAUNCH VEHICLES			
Delta II 7920/25	3,275	50	47/49
Delta II 7320	1,750	35	0/0
Pegasus XL	225	14	18[a]/23[b]
Taurus XL/Orion 38	945	24	0/0
Taurus/Orion 38	860	22	3/3
Athena 3	2,200	30	0/0
Athena 2	700	22	1/1
Athena 1	200	16	1/2
Conestoga 1229	220	12	0/0
Conestoga 1620	540	18	0/1
FOREIGN LAUNCH VEHICLES			
CZ-2D (China)	1,200	20	5/5
PSLV Mk2 (India)	1,300	12/15	1/1
Molniya M (Russia)	1,775	30	256/289
Shavit 2[c] (Israel/US)	340	15	0/0
Shtil 1N (Russia)	185	5/6	1/1
Tsyklon 3 (Ukraine)	2,300	25	111/117

[a] Successes exclude incorrect orbit, failure to separate on orbit, and damaged spacecraft.
[b] Includes all versions of the Pegasus.
[c] Coleman Research Corporation, in collaboration with Israel Aircraft Industries, has recently won a Small Expendable Launch Vehicle contract from the National Aeronautics and Space Administration to provide launch services in the United States using an export version of the Israeli-designed Shavit rocket (Next). The Shavit is a solid-fuel rocket with performance comparable to the Pegasus XL. Through January 1998, it had achieved three successful launches in five attempts.
SOURCE: International Space Industry Report, Nov. 9, 1998; available online at <CS:WebLink>http://www.launchspace.com/isir/home.html>.

Table 5.2 Launch Vehicle Fairing Dimensions and Launch Environment

	Pegasus XL	Athena 1 (Mod 92 fairing	Taurus (63 in fairing	Conestoga (1229 fairings)	Delta II (7920 (9.5 ft fairing)
Fairing dimensions					
Max diameter (m)	1.118	1.981	1.372	1.616	<2.54
Max cylinder length (m)	1.110	2.291	2.692	0.392–2.664	3.81
Max cone length (m)	1.016	2.002	1.270	1.768	1.94
Launch environment					
Axial accel (g)	<13.0	<+4/-8	<11.0	<11.0	<6.0
Lateral accel (g)	<±6.0	<±2.5	NA	<±2.7	<±2.0
Acoustic (dB)	<141	<133.5	<141	<128.5	<139.6
Longitudinal freq (Hz)	NA	>15	NA	NA	>35
Lateral freq (Hz)	NA	>30	NA	NA	>15

NA = not applicable.

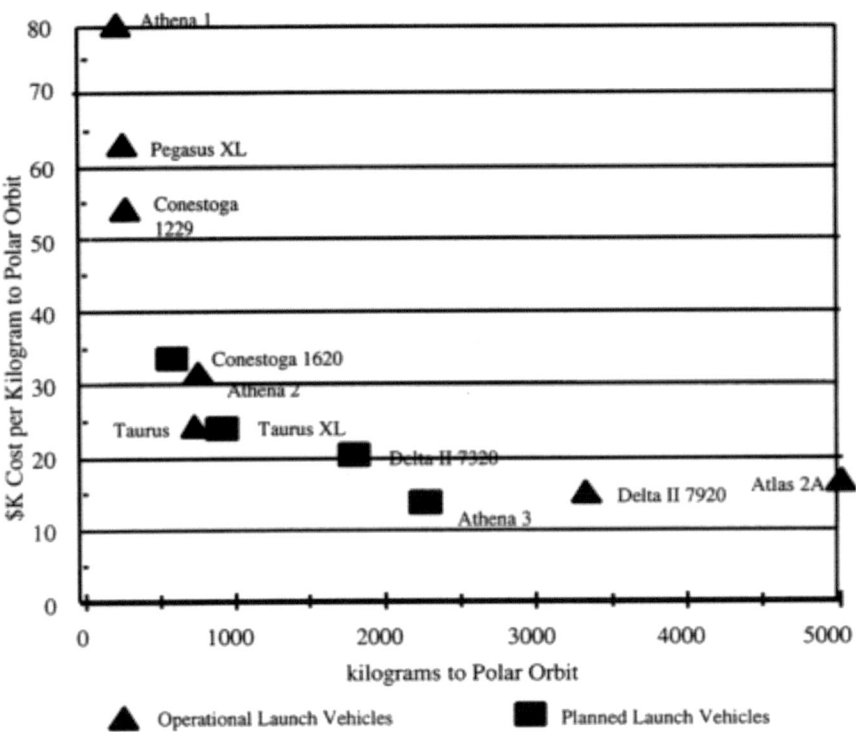

Figure 5.1
Launch vehicle cost per unit mass to EOS orbit.

SUMMARY

Achieving the full promise of small satellites will require the availability of reliable U.S. launch vehicles with a full range of performance capabilities. This is currently not the case: There is a significant gap in capability between the Pegasus/Athena/Taurus launch vehicles and the Delta II. Plans to fill this gap by numerous suppliers are encouraging, as are the efforts by launch vehicle suppliers to provide a range of fairing sizes to accommodate a larger percentage of potential missions. Foreign launch vehicles may also ultimately play a role in filling this gap, should U.S. policy change.

Early experience with the new small launch vehicles has included a number of failures—probably due in part to a desire to minimize development costs for these commercial ventures. Continued development should overcome the difficulties and yield a suitable balance between cost and reliability. However, it will take some time—and, likely, some additional failures—before any of these launch vehicles establish a reliability record approaching that of the Delta II.

6

Small Satellites and Mission Architectures

OPTIONS FOR DISTRIBUTING SENSORS

A single-sensor satellite mission clearly calls for the smallest spacecraft bus that can accommodate the payload and the smallest launch vehicle that can deliver the satellite to the desired orbit, as this will be the most efficient and lowest cost solution. The right choice of launch vehicle is less obvious when a mission involves multiple sensors and options are available for the number of satellites, their size, and how to aggregate or distribute the sensors among them. The choices for such a mission include single-sensor platforms, multisensor platforms (e.g., Earth Observing System [EOS] AM-1, Polar-orbiting Operational Environmental Satellite [POES]), clusters, and constellations. *Clusters* refer to a collection of two or more satellites relatively closely spaced in a common orbit (formation flying). *Constellations* refer to a collection of satellites whose relative positions are controlled in each of multiple orbits; examples include the Global Positioning System (GPS) and the Iridium communication satellites.

Single-Sensor Platforms

Single-sensor platforms provide the greatest mission flexibility. Technically, they allow for unique orbits for each sensor, they avoid interference among sensors on a common platform, and they permit trade-offs between using the spacecraft bus itself versus additional sensor mechanisms to perform calibration or observing maneuvers. Programmatically, their relative simplicity makes them less subject to the schedule delays and budget impacts a late sensor delivery can have on multisensor programs. They typically require shorter mission development times (24 to 36 months) and can thus employ more current technology and deliver faster "time to science." A complement of sensors on multiple small satellites can be launched sequentially as budgets and schedules permit; and, in the case of a failed sensor, a direct replacement satellite can be launched without having to deal with residual assets on a multisensor platform.[1] On the negative side, as discussed in Chapter 5, the U.S. fleet of small expendable launch vehicles suitable for small satellites (e.g., the Pegasus, Athena, Conestoga, and Taurus) are relatively unreliable at present, and the probability of one or more launch failures where multiple launches are

---

[1] If the mission requires that data from all sensors be available contemporaneously, sequential launches and development delays in any of the single platforms will of course affect the mission. Delays in single platforms can also be troublesome when establishing certain kinds of constellations—for example, the Iridium communication network, which relays signals between satellites distributed in low-Earth orbit.

required is significant. Also, mission operations grow more complex and costly as the number of satellites needed to perform a mission increases.

Multisensor Platforms

Requirements for spatial and temporal simultaneity among measurements have traditionally led to the use of multisensor platforms. Placing multiple sensors on a common platform ensures contemporaneous measurements and permits coalignment such that they view the same scene simultaneously. This is always desirable when the measurements are complementary and is in some cases essential if full value of the data is to be realized.

While single-sensor platforms provide the greatest mission flexibility, multisensor platforms have a higher probability of delivering an equivalent number of sensors to orbit because they require fewer launches and use more reliable launch vehicles (e.g., Delta, Atlas).[2] Multisensor platforms also offer simpler ground system operations with fewer spacecraft to control and fewer data downlinks.[3] On the other hand, design compromises may be needed because all sensors must have a common orbit, and payload accommodations must account for possible optical (fields-of-view), mechanical (jitter), or electromagnetic sensor interference as well as the possible multiplexing of resources (data handling, power, etc.). Further, with a multisensor platform, a launch or satellite bus failure results in the loss of all sensors.

Clusters

When spatial or temporal simultaneity of measurements is required, flying several single-(or few) sensor spacecraft in formation as a cluster is an alternative to a larger multisensor platform. This addresses the requirements for simultaneity of measurements while still providing the programmatic flexibility associated with small spacecraft: For example, the cluster can be built up incrementally, and, in the case of failure or obsolescence, sensors can be replaced one (or a few) at a time.

Formation flying is feasible but has not been proven in operational use.[4] The difficulty and cost associated with this approach depend on how stringent the requirements are for coalignment and simultaneity of measurements. In all cases, on-board propulsion systems are needed to maintain relative position within the cluster. Coalignment of sensors on separate platforms to better than a few tenths of a degree requires sophisticated spacecraft attitude control and/or sensor pointing systems. In addition, active control links between spacecraft are

---

[2] A spate of launch vehicle failures occurred in the months before this report was completed, including failures on larger and historically more reliable launch vehicles (Delta, Titan) as well as newer launch vehicles (Athena). On May 4, 1999, the second stage of a Delta III rocket failed, leaving a $150 million satellite in a useless orbit. On April 30, 1999, a Titan IV rocket failed to place an $800 million Milstar communications satellite into its proper orbit. On April 27, 1999 a Lockheed Martin Athena 2 rocket failed to place a commercial space imaging satellite into polar orbit. On April 9, 1999, a $250 million missile warning satellite was left stranded in the wrong orbit after its upper stage booster failed on a Titan IV rocket. On August 26, 1998, a Delta III exploded on its maiden flight, destroying a communications satellite. On August 12, 1998, a Titan IV rocket carrying a $1 billion classified payload exploded shortly after launch. See the extended launch and explore the archives pages at FLORIDA TODAY Space Online (1999).

[3] The size, complexity, and cost of a satellite ground system are important issues when there are potentially many satellites in the same orbit and ground contacts overlap. The National Oceanic and Atmospheric Administration and the Air Force typically retire (essentially through lack of attention) older polar-orbiting meteorological satellites because of the load on their tracking and processing facilities. In addition, generation of command lists, ephemerides, and the other data required to operate a satellite requires personnel, time, and money. To better manage these problems, operation of multisatellite systems requires approaches to operations that are more automated and less taxing of human resources.

[4] Earth Orbiter-1 (EO-1) is the first of a series of Earth orbiting missions for the National Aeronautics and Space Administration's New Millennium Program. The mission, which is scheduled for launch on April 13, 2000, is designed to validate a number of technologies that would provide Land Remote Sensing Satellite (Landsat) follow-on instruments with increased performance at lower cost. The mission's centerpiece is an Advanced Land Imager (ALI) instrument. Once on orbit, EO-I will provide 100 to 200 paired scene comparisons between ALI and the Landsat 7 imager, ETM+ (Enhanced Thematic Mapper Plus), to validate ALI's novel multispectral technology. The mission will also demonstrate formation flying, since the EO-orbit will be associated with that of Landsat 7. The EO-1 spacecraft will fly in a Sun-synchronous orbit at the same altitude as, but slightly offset from, that of Landsat 7. Thus, EO-1 will fly over the same ground track as Landsat 7, but several minutes behind it. For more information on the mission, see EO-1 (1999).

needed for close control of formation spacing. A tight formation leads to data communication challenges, because it requires multiple ground antennas or relay links between satellites. Formations with widely spaced satellites allow individual satellite contacts to be accomplished with a single ground antenna (15-minute separation) and therefore are easier to manage at relatively modest cost. The GPS is an example of such a constellation. For Earth remote sensing, such a constellation may be spaced too far apart for adequate sensor data registration.

Constellations

For missions requiring global or continuous coverage, there is a potential advantage in deploying a constellation of spacecraft with similar sensor(s) in one or more orbits. In fact, there may be no alternative to this approach if global coverage must be provided in real time or within a very short interval. Systems such as the GPS or the newer mobile telephone systems need multiple orbits and satellites to provide the requisite coverage. Measurements of some Earth system processes may not require near-continuous coverage, and the number of orbits and satellites may be reduced at the expense of increased delays in data acquisition. Such constellation architectures are robust; they also increase flexibility in the overall observing program. The constellation can be built up incrementally using multiple or single satellite launches. If the constellation is composed of many satellites, the overall system is relatively insensitive to the failure of an individual satellite. Replacements can be made one at a time as they are needed. If there are few satellites in the constellation, or if it is critical that the data be collected all the time, then the constellation can include spare satellites on-orbit.

Depending on the mission, multisatellite constellations can involve complex communications links and/or ground operations. To date, constellations have been applied, or studied for application, in operational missions involving communications (Iridium, Globalstar, Odyssey, Teledesic); navigation (GPS); and surveillance (Space Missile Tracking System).

Constellations of small satellites may also offer a new approach to those science missions that would benefit from more frequent sampling by a larger number of lower cost (and presumably lower accuracy) sensors. Monitoring of time-varying phenomena such as cloud cover might be a good example of a mission where higher refresh rate at lower accuracy is a preferred approach.

The relatively high cost of fielding multisatellite systems may limit their potential for dedicated research missions even though the use of many identical spacecraft offers economies of scale in producing space segment hardware. Accommodating research sensors as additional payloads on operational satellite constellations could offer an effective approach to reducing cost.

COST-EFFECTIVENESS OF SMALL SATELLITE ARCHITECTURES

One of the key factors driving the strong current interest in small satellites is a desire to reduce mission costs, largely as a result of very severe budget pressures. When discussing small satellite mission costs, it is important to distinguish between small (low-cost) missions and larger missions that are performed with multiple small satellites. There is little question that valuable missions can be accomplished at relatively low cost with smaller payloads, satellites, and launch vehicles. This is the philosophy currently employed by the National Aeronautics and Space Administration in selecting missions for the Discovery, Small Explorer, Mid-size Explorer, and Earth Science System Pathfinder programs. In all of these programs, cost maximums are defined as a mission design requirement, and only those missions that can be performed under the caps are considered for selection. Because smaller systems are typically less costly than larger ones, this has led to a series of small satellite missions.

A different situation exists when a mission is defined in terms of the needed measurements or existing payloads, and the objective is to select the best system architecture to accomplish it. Cost is once again a key parameter, but here it is a criterion for comparing alternative architectures rather than a constraint on a mission's definition. Where it is scientifically appropriate, a key question in this context is thus whether it is cost-effective to split up a large multisensor payload onto multiple smaller satellites. Because alternative architectures involve varying numbers of spacecraft, launches, and ground station elements, the specific cost per pound to place, maintain, and operate a specified payload in orbit is the parameter of interest rather than the absolute cost of individual system elements.

As discussed in Chapter 4, a recent RAND study (Sarsfield, 1997) indicates that, on a specific cost (per unit mass) basis, small satellite development costs can be relatively high when they retain the complexity needed to achieve demanding scientific requirements (see Figure 4.1). Further, as shown in Chapter 5, launch costs per pound of payload (satellite) increase substantially for smaller launch vehicles. Consequently, the cost of delivering a given set of sensors to orbit will be greater on multiple small satellites than on a multisensor platform.[5] It is also clear that operational costs will be higher for a larger number of satellites, including the costs for communications, command, and control, as well as data processing when registration correction is required. However, the cost of delivering sensors to orbit is only part of the total cost associated with a remote sensing program. A key issue—and one that can drive the architectural cost trade-off—is the cost of maintaining a particular remote sensing capability. This is discussed in the following section.

Maintenance

Maintenance can be a driving cost factor depending on the mission success criteria. For an operational mission (e.g., POES), the availability[6] requirement usually means that loss of a key sensor requires launch of a replacement. However, the POES satellites are multisensor systems. Replacement of a key failed sensor on POES, or on the National Polar-orbiting Operational Environmental Satellite System (NPOESS), should its satellites also be multisensor systems, can result in a mix of fully operational multisensor satellites and residual operational assets from the failed satellite. These residual assets can be factored from the start into mission planning to minimize future replacement requirements, but at some cost in operational complexity (operating more spacecraft). Employing single-sensor satellites to replace failed sensors on multisensor platforms avoids creating residual assets but involves similar operational complexities—and, if sustained as a strategy, ultimately leads to a complete single-sensor platform architecture. A superior strategy might be one in which the decision to launch single-sensor or multisensor platforms to replace failed sensors is based on the depreciated value (i.e., expected remaining life) of the still operating sensors.

Scientific missions seldom have an availability requirement for a particular sensor(s), although some missions are deemed important enough that replacement occurs after a failure. Typically, loss of a sensor results in loss of the data stream (and not a higher mission cost to replace the sensor). No residual assets are created by replacement, and a multisensor platform continues to function as long as useful data are generated and operations are affordable.

Thus, the implications of a failure—and hence the trade-off between smaller and larger platforms—is quite different for the operational National Oceanic and Atmospheric Administration missions (e.g., POES, the Geostationary Operational Environmental Satellites [GOES], and, in the future, NPOESS) versus the research-oriented EOS program. This is illustrated in the analysis presented below of the cost-effectiveness of alternative mission architectures for NPOESS.

NPOESS

Both the Johns Hopkins University Applied Physics Laboratory (APL) and TRW have performed NPOESS architecture studies. At first glance, the studies appear to have arrived at opposite conclusions regarding the potential for cost reduction by accommodating the multiple NPOESS sensors on several small satellites versus a single larger one. However, further analysis showed that both results were consistent with the assumptions used and that the preferred architecture is highly dependent on those assumptions (Raney et al., 1995; Rasmussen and Tsugawa, 1997). A summary of the key reliability assumptions used for these studies is presented in Table 6.1.

---

[5] The lower cost of individual small satellites and small launch vehicles is more than offset by the need for multiple units.

[6] Availability refers to the percentage of time the system is able to deliver acceptable data from critical sensors to the users. *Instantaneous availability* refers to the probability that the system can deliver the data at any given time; *average system availability* refers to the probability that the system can deliver the data averaged over the required mission life.

Table 6.1 Reliability Assumptions Used in NPOESS Mission Architecture Studies

Mission Component	Raney et al. (APL)	Rasmussen and Tsugawa (TRW)
Launch vehicle		
Large	0.95	0.95
Medium	0.95	0.90
Small	0.95	0.90
Critical sensors	0.80 at 3 years	0.84 at 7 years
Spacecraft bus	0.96 at 3 years	0.89 at 7 years

To explore operational mission cost sensitivity to satellite size, Rasmussen and Tsugawa considered large, medium, and small satellite classes sized to the mass and volume constraints of the Delta, Lockheed-Martin Launch Vehicle-3 (LMLV-3), and Taurus launch vehicles. Three separate architectures were constructed using one large, three medium, and six small spacecraft in each of two orbit planes such that each architecture carried the same set of 10 sensors per orbit (4 critical, 6 noncritical) and had the same mission objectives. The primary mission objective was to achieve specified space segment availability for critical sensors over the life of the mission. Space segment and launch costs, as well as the ability to meet mission objectives, were used to judge the viability of each architecture. Results were obtained as a function of system parameters such as sensor and spacecraft reliability and design life, relative spacecraft costs, and the desired system availability.[7]

Figure 6.1 illustrates some key findings of this study . The figure shows the sensitivity of space segment mission (life cycle) costs to sensor reliability for a 10-year mission.[8] Results are shown for the large, medium, and small satellite architectures where the individual recurring bus costs are realistically estimated to be in the ratio 1:1/2:1/4 and the average system availability requirement is >90 percent.

A key study finding was that small, medium, and large satellite architectures have different life cycle cost sensitivities to sensor reliability. As a result, the most cost-effective architecture is sometimes a small satellite architecture and sometimes a large satellite architecture. Large satellite architecture costs are more sensitive to sensor reliability because they carry more sensors, all of which are replenished when a new satellite is launched in response to any critical sensor failure. When sensor reliability is high and failure infrequent, the lower cost to deploy the payload with fewer, larger platforms outweighs the added costs of occasionally launching unnecessary sensors and provides a life cycle cost advantage to large satellite architectures. On the other hand, low sensor reliability, with concomitant frequent replacement, leads to excessive unnecessary sensor replacement with large platforms and favors small satellite architectures.

This behavior explains the disparate results between the APL and TRW studies. The APL study assumed low reliability, short-lived sensors typical of those flying on POES today. For this class of sensor, Figure 6.1 shows that the results favor the small satellite architecture. The TRW study assumed higher reliability and longer design life, reflecting the requirements being used to develop the next generation of operational sensors for NPOESS. These assumptions favor the larger, Delta class architecture and lead to substantially reduced mission costs.

Availability is another key parameter affecting mission costs for operational systems. Each system has a cost and average availability determined by the reliability and design lives assumed for the sensors and spacecraft bus,

---

[7] This study did not include the ground segment in mission cost trade-offs. It is likely that ground segment costs increase with a larger number of satellites in orbit and that the trends in the study results, which show life cycle cost advantages for multisensor satellites, would be sustained and possibly amplified by including them. This should be verified with further study.

[8] Mission costs are presented as a ratio normalized to the large (Delta class) satellite architecture solution (with the Rasmussen and Tsugawa reliability assumptions of Table 6.1). Mission costs are similarly normalized with respect to the TRW Delta class architecture solution in Figures 6.2 and 6.3.

the launch vehicle reliability, and the assumptions made regarding replenishment criteria and launch schedule. For the current two-satellite POES system, new satellites are theoretically launched on demand, that is, within 120 days upon failure of a critical sensor or a spacecraft bus. Average availability of only 80 to 90 percent for two operational satellites can be achieved with this approach (during the time between failure and replacement, the data product is not available). To achieve higher average availability, new replacement satellites must be launched periodically—on a schedule—in addition to launching on demand to accommodate random failures. The sensitivity of mission life cycle cost to average availability for this scenario (using the Rasmussen and Tsugawa assumptions of Table 6.1) is shown in Figure 6.2 for the small, medium, and large satellite architectures studied.

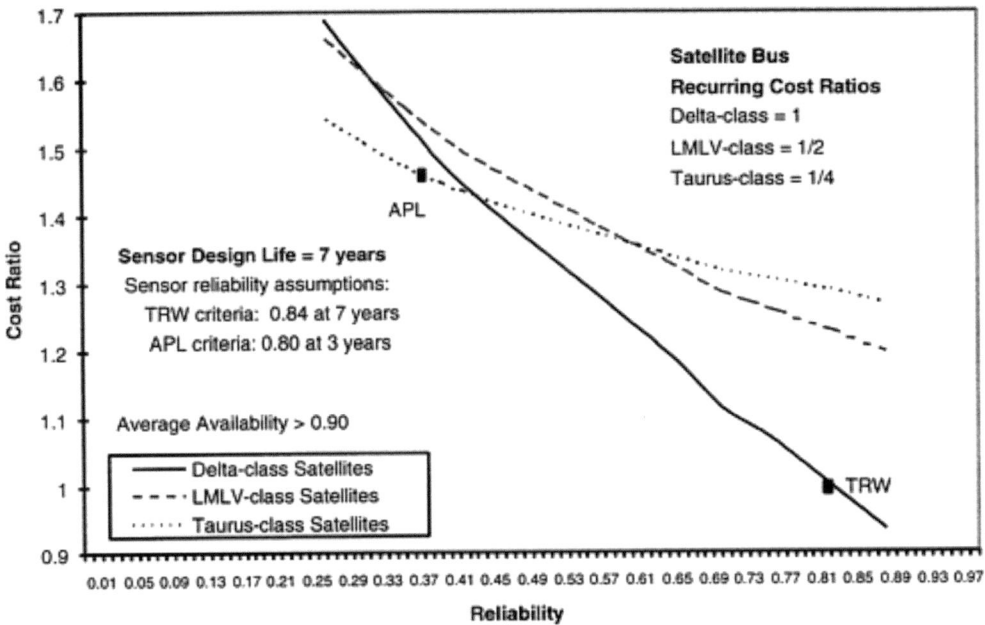

Figure 6.1
Relationship between mission cost and sensor reliability: operational mission model.

For each architecture, relative cost versus mission average availability is plotted for scheduled replacement launches every 6, 5, 4, 3, and 2 years (in addition to launching in response to random failure). Average availability in the range of 94 to 97 percent is achieved with scheduled launches every 6 years. Achieving higher average availability by launching replacement satellites more frequently involves substantially higher costs. Clearly, mission requirements for very high availability should be scrutinized carefully.

Earlier, it was demonstrated that the cost to deliver a given set of sensors to orbit favored the use of multisensor satellites. Here we see that the cost versus availability results again favor larger satellite architectures because of the sensor and satellite bus reliability assumptions and fewer anticipated launch failures.

EOS

Nonoperational, Earth science missions that require multiple sensors on orbit such as EOS AM, PM, and Chemistry are also subject to trade-off studies regarding the number and size of satellites deployed and the parsing of sensors. As with operational missions, results depend on the assumptions used and the criteria for success. To gain insight into this class of missions, Rasmussen and Tsugawa (1997) studied a generic 5-year science mission in which four sensors are launched into one or more orbits on either one large, two medium, or four small satellites, again consistent with the launch capabilities of the Delta, LMLV-3, and Taurus vehicles. Typical of science missions, no replenishment launches were assumed regardless of failure of launch vehicle, spacecraft bus, or sensors.

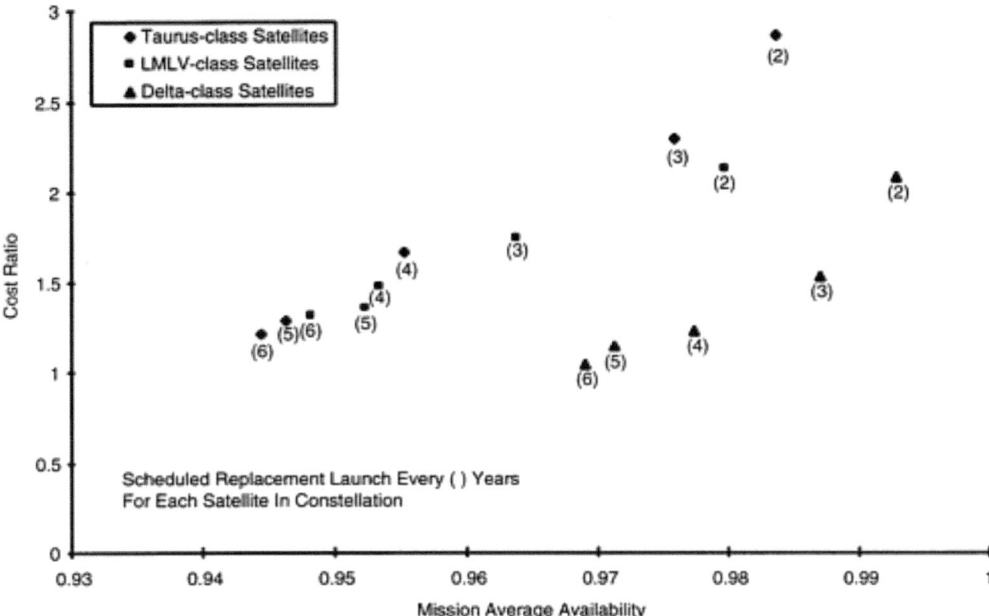

Figure 6.2
Mission cost sensitivity to average availability: operational mission model.

These assumptions lead to a straightforward cost comparison based on the estimated costs to develop and launch the various satellites that make up the initial architectures. Figure 6.3 shows the relative mission costs, using the same element cost assumptions as the operational weather system (POES), and again normalized to the single, Delta class satellite case. Parsing a given complement of sensors among a larger number of independently launched satellites increases mission cost. A recent NASA study on alternative architectures for the EOS Chemistry-1 mission led to similar results.[9]

But costs alone do not define the preferred solution; they must be viewed in terms of the success criteria for the mission. For a science program, a reasonable criterion for success might be related to the number of sensors still operating at the end of the mission. The probability of launching and sustaining a completely operational system (four working sensors) as a function of time over the 5-year mission life is shown in Figure 6.4. Multisatellite architectures are clearly penalized by the greater probability of at least one launch vehicle failure as reflected in the initial points for each curve.

---

[9] The NASA study is discussed in EOS (1996).

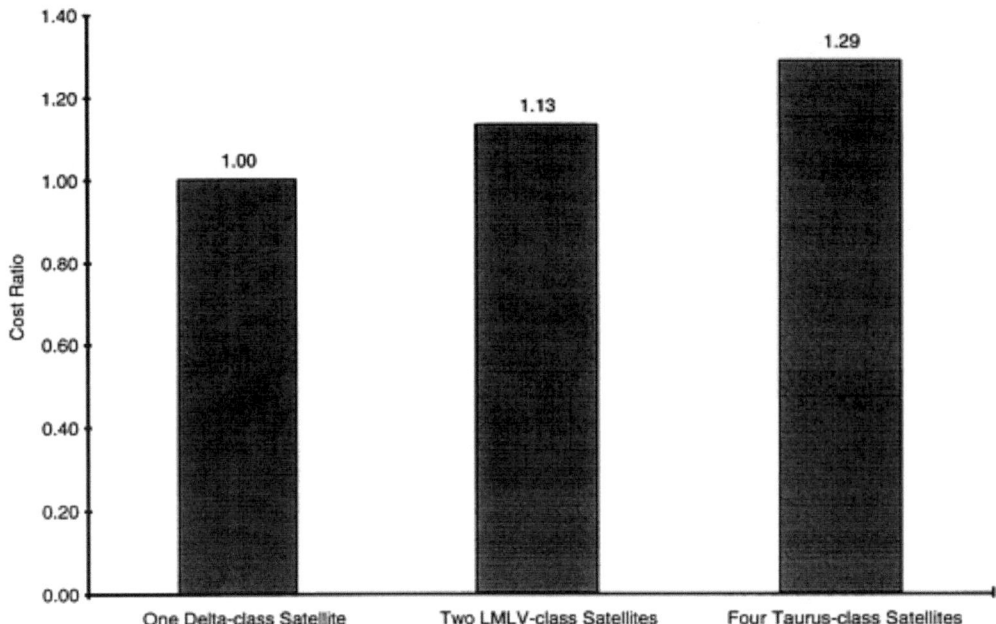

Figure 6.3
Relative mission architecture costs: science mission model.

The analysis above changes, however, if limited failures are permitted within the context of a successful mission. The probability of completing the 5-year mission with a specific number of failures is shown in Figure 6.5 for each architecture. If mission success requires that at least three of the four sensors be operational over the full 5 years, then the single Delta class satellite is the most effective architecture in terms of having the highest probability of success. If only one or two surviving sensors (three or two failures) can still be considered a success, then the medium and small satellite architectures are more effective solutions. The highest probability case occurs when only one of four single-sensor small satellites needs to survive the launch and 5 years of operation for the mission to be considered successful. Thus, the optimal architecture for scientific multisensor missions is dependent on the success criteria established. The explicit requirements must be examined in order to select the correct system architecture to perform a mission in the most cost-effective manner at an acceptable level of risk.

Finally, it should be recognized that the preceding analysis has not considered such qualitative factors as the "time to science"; that is, the time it takes to develop a mission and deliver scientifically useful data. For some missions, particularly those that are research-oriented, the potential to answer pressing scientific questions more rapidly may override other considerations in justifying a faster, more flexible small satellite solution.

SUMMARY

Small satellites—in particular single-sensor platforms—provide great architectural and programmatic flexibility. They offer attractive features with respect to satellite design (distribution of functions between sensor and bus); observing strategy (tailored orbits, clusters, constellations); time to science for new sensors; replenishment of individual failed sensors; and robustness with regard to budget and schedule uncertainties. On the other hand, spatial and temporal simultaneity of measurements are easiest to achieve with multisensor platforms. And, because fewer launches are involved, multisensor platforms offer a higher probability of placing a given complement of sensors into orbit without loss. They also frequently offer the simplest ground segment solutions, including mission operations, downlink and data system architectures, and calibration and validation of sensors. Clusters or constellations of small satellites may require additional ground station elements or spacecraft data crosslinks, in addition to more challenging mission operations planning, to accommodate increased communication, command, and control requirements.

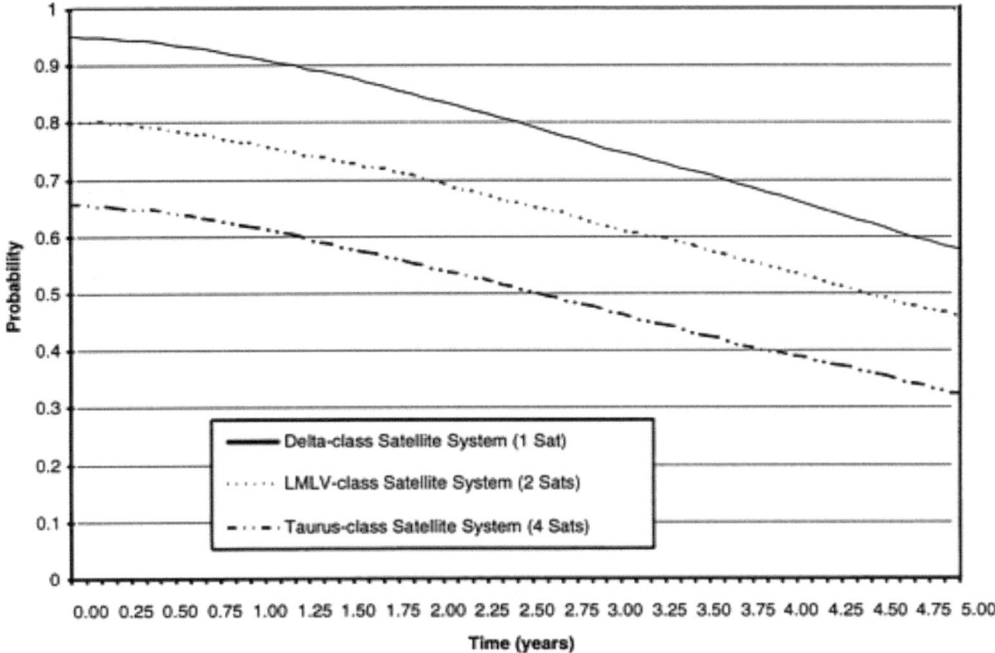

Figure 6.4
Probability of operation with no failures versus time: science mission model.

Cost trade-offs for single-versus multisensor platforms and for single-versus multisatellite architectures are driven by the costs, reliabilities, and design lives of the system elements (sensors, spacecraft buses, launch vehicles, ground segments) and by availability requirements for operational systems. The following observations generally hold true:

- The lowest cost to place a given set of sensors into orbit will often be with the smallest suitable multisensor platform.
- The lowest cost architecture to maintain a set of operational sensors in orbit for a sustained mission life is mission specific and must be determined on a case-by-case basis.
- Small satellites may provide economic benefits as part of a replacement strategy for failed sensors or for sensors with limited design life or reliability.

Differences in success criteria between research and operational missions affect mission architecture trade-offs. Life cycle costs and availability requirements drive decisions regarding the configuration of operational missions, whereas the urgency with which data are needed (time to science), tolerance to partial failure, and programmatic flexibility are factors that may override cost considerations in certain research missions.

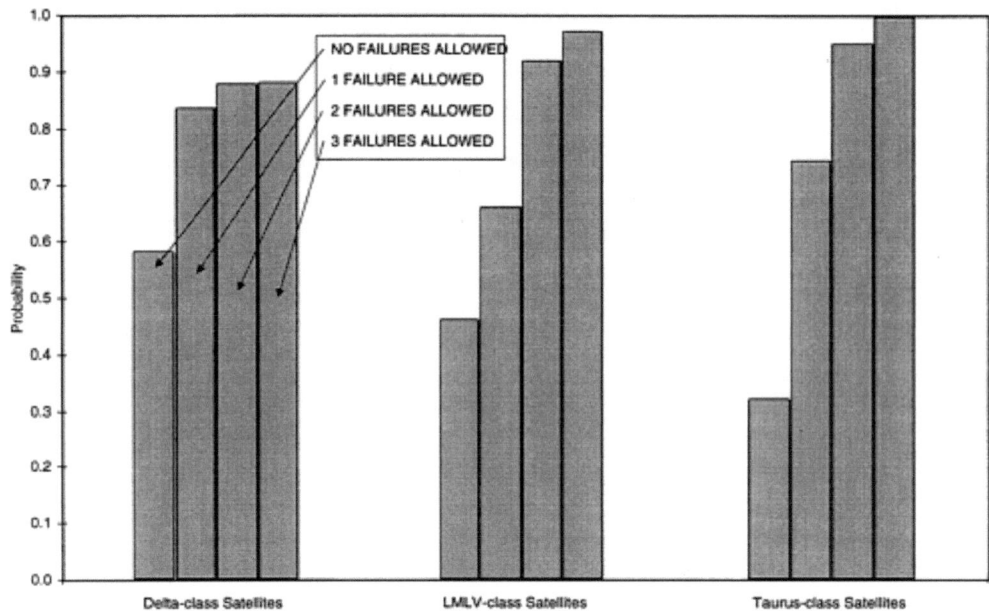

Figure 6.5
Probability of mission success defined in terms of a specific number of failures: science mission model.

REFERENCES

Earth Observing System Project Office (EOS). 1996. Report of the EOS Payload Panel on the GSFC CHEM-1 study. The Earth Observer 8(3). Available online at <http://eospso.gsfc.nasa.gov/eos_observ/5_6_96/p15.html>.
Earth Orbiter-1 (EO-1). 1999. Available online at <http://fdd.gsfc.nasa.gov/missions/eo-1.html>.
FLORIDA TODAY Space Online. 1999. Available online at <http://www.flatoday.com/space/>.
Raney, R.K., G.H. Fountain, E.J. Hoffman, P.F. Bythrow, and R.H. Maurer. 1995. Small Satellites and NOAA: A Technology Study. The Johns Hopkins University Applied Physics Laboratory, American Institute of Aeronautics and Astronautics/Utah State University Conference on Small Satellites, Utah State University, Logan.
Rasmussen, A., and R. Tsugawa. 1997. Cost Effective Applications of Constellation Architectures of Large, Medium and Small Satellites. AIAA Paper 97-3950. Reston, Va.: American Institute of Aeronautics and Astronautics.
Sarsfield, L. 1997. The Cosmos on a Shoe String: Small Spacecraft for Space and Earth Science. MR-864-OSTP. Santa Monica, Calif.: RAND, Critical Technologies Institute.

7

Opportunities and Challenges in Managing Small Satellite Systems

With few exceptions, small satellite programs are currently implemented using traditional management approaches. However, realizing many of the potential benefits of a small satellite approach requires innovations in programmatic style. That is, as with larger satellites, small satellite programs can be costly or slow to implement. Similarly, within some limit, larger satellite programs can be managed in ways that result in comparatively low-cost, short-development-time missions. (Clearly, very large and complex missions require more time for development and integration.) The discussion below examines some of the management and programmatic issues related to small satellite missions and presents committee views on how to reap the potential benefits of a small satellite approach.

PROGRAMMATIC APPROACHES TO TECHNICAL ISSUES

The fundamental attribute of small satellite missions is their potential to shorten development time; all other benefits flow from this basic property. With longer development cycles, total system costs increase. Technological advances take place, but they cannot be pursued because the design is completed long before the mission is scheduled for launch. Improvements in scientific understanding that may occur during the development of the mission often have little impact on sensor design.

There are many ways to achieve shorter development times through changes in management practices. One way is for the government to limit the oversight burden by reducing the number of documents and reports that must be delivered by the contractor and the number of formal reviews that must be undertaken. For example, it has been estimated that a single major review of the Earth Observing System Data and Information System (EOSDIS) core system being built for the National Aeronautics and Space Administration (NASA) produced over 5,000 pages of material at a cost of several person-years of effort.[1] TRW estimates that the Critical Design Review for the Total Ozone Mapping Sensor, now in orbit as part of the Earth Probes program, required 10 person-years of effort. A lower cost approach is to reduce the required paperwork to sufficient information to allow a knowledgeable engineer or manager to evaluate progress and risks. Such an approach has been dubbed "insight versus oversight."[2]

---

[1] Bruce Barkstrom, NASA Langley Research Center, personal communication.
[2] Strategies to reduce mission development time are discussed in NRC (1997a,b).

Although there is anecdotal evidence of too much required paperwork, the committee is not aware of any formal study of optimal levels of oversight.[3] In any event, reduction of oversight does not mitigate the importance, or lessen the involvement, of government personnel, as they play a valuable role in ensuring mission success if they are a working part of the team rather than just project observers.

Many programs—Earth Observing System (EOS) AM is an example[4]—have incorporated NASA personnel into the product development teams. The Air Force has also been well integrated into several programs that used the product team approach. However, it is difficult to incorporate these plans into a proposal, as the government's evaluation of cost and effort can be adversely affected if it is believed that the contractor intends to use the government personnel to make decisions or accomplish tasks that should be done by the contractor. Closer integration between the technical personnel and the end user could also achieve shorter development times. This approach conceivably could extend into the proposal phase as well as the development phase. In this way, there would be fewer breaks in communication between producers and users, and new developments on either the technical or scientific side could be rapidly incorporated into the mission.

Building on experiences in the commercial manufacturing industry, contractors could rely more heavily on integrated product development teams rather than dividing the effort, as is typically done, into separate mission components. This approach could extend to all parts of the mission, including the ground, launch, bus, and sensor systems and might reduce failures due to lack of communication as well as stimulate more creativity and innovation. Notable examples of this approach can be seen in the automobile industry and in much of the Japanese manufacturing industry where individual companies regularly work together to produce a complete system. The General Motors Saturn automobile plant follows this model, as worker teams participate in the end-to-end manufacturing process.[5]

RISKS

Programmatic Risks

There are many risks to the management approaches outlined above. Reduced government oversight may substantially increase the risk of failure by not providing an outside viewpoint during the development process. Shorter development times (and perhaps lower profit margins) may encourage companies to take shortcuts and higher risk (but lower cost) options without a thorough analysis. None of these potential problems is an expected or proven consequence of reduced oversight; indeed, some might occasionally occur with present levels of government oversight.

Satellite development and launch operations are complicated, and seemingly innocuous errors can propagate through the system. For example, one of the early Pegasus XL missions failed because designers relied only on numerical modeling of vehicle aerodynamics rather than actual wind tunnel tests. When the model was run with incorrect numbers, the resultant control software became flawed. Fleeter (1997) reports that a similar reliance on models rather than actual tests resulted in the near failure of the Clementine mission.[6]

Government procurements are often based on rapid analyses by contractors. This leads to the additional risk of not detecting errors that might otherwise be found in a longer government review. In the best scenario, these errors are discovered when the actual development effort begins. Costs may escalate, but the program is not compromised. In a more pessimistic scenario, the errors are not found and program failure may result.

---

[3] An informal review of software development deliverables was done for the Space Station Program in approximately 1994; it showed that the Institute of Electrical and Electronics Engineers required 4 documents; the Air Force, 18; and the Goddard Space Flight Center, 21. The NASA requirements have since been reduced, but not drastically. A Department of Defense initiative to reduce the use of imposed standards has been reflected in requests for proposals since about 1996 and has reduced some of the agency's formalism of control.

[4] L.C. Scholz, personal communication, based on experience at Lockheed Martin Missiles and Space on the EOS program.

[5] See, for example, "Integrating Skill and Experience on the Shop Floor" in Shaiken et al. (1996).

[6] A Clementine control system thruster had been incorrectly designed so that it would turn on, but not off. As a result, the spacecraft spun until the errant thruster ran out of fuel. Clementine's failure may actually have been due to a problem in software that was loaded after the primary mission was completed. The point of the discussion is that success or failure can occur with or without oversight.

Current government funding and procurement practices may be out of step with the style of proposal and system development best suited to small satellite missions. For example, detailed procurement rules may significantly delay acquisition of critical parts or may force contractors to acquire parts much earlier than needed. Funding profiles may not reflect the need for rapid, early delivery of funds necessary to support a short development schedule. Requirements for long-lead-time space-qualified parts may actually decrease mission reliability by compressing the time available for assembly, integration, and test.

More subtle issues arise on the contractor side. Integrated development teams may require that several companies work closely together on the overall system. Such close cooperation depends on open communication, but this will be difficult if concerns over proprietary information are present. The U.S. space industry is not presently structured as a set of independent component builders. Few companies are involved in only one aspect of satellite missions, and many have business units that could provide end-to-end solutions from launcher to bus to sensor. Thus companies may be unwilling to disclose proprietary information to a partner that may be a potential competitor on a future project. Given industry trends of reduced profit margins and increased competition, these concerns are likely to increase.

The computer industry is often cited as an example of the benefits of a "faster, better, cheaper" approach. It is notable, however, that few companies manufacture all of the major subsystems of a computer—central processing unit (CPU), disk, backplane, operating system, and applications software. Thus a chip manufacturer such as Intel is more willing to work with an operating system vendor such as Microsoft on the details of a new CPU because it knows Microsoft will not be competing in the CPU marketplace. Tighter alliances between competitors in the space industry may run counter to traditional business practices as well as to government regulations.

Management of Programmatic Risks

Desirable characteristics of small satellite missions, such as more rapid development schedules and lower costs, are also associated with an increased likelihood of mission failure. Understanding, mitigating, and responding to the risk of failure are thus central issues in small satellite programs.

Space launch and space operations are inherently risky. One response to failure is to launch another mission immediately. The formerly classified Corona program of photo-reconnaissance missions in the late 1950s and 1960s was based on this principle. It took 12 failures before there was a successful mission (Ruffner, 1995). This strategy must be based on a firm scientific and political recognition of the vital importance of the specific mission. Other failures may be less dramatic. A contractor may simply run out of funds during development or may fall far behind schedule. In such cases, the response may be to cancel the mission outright (which has rarely been done, but see the discussion of the Clark mission in Appendix D) or to increase funding in the hope that the program will recover.

Recent government practice has been to encourage contractors to deliver within a fixed cost. The government perception is that larger companies more easily absorb cost overruns. This further reduces profit margins and makes it difficult to participate in such contracts, because the profits are also limited on the upside of a successful program. Such risks must be justified to corporate investors, who may be unwilling to risk the downside potential of such investments. The net effect may be to filter larger companies out of the process, creating a reliance on small companies (which have limited capital) to participate in small missions.

If the government is not willing to accept failures in the development process, then it must be prepared to either manage the risks in this process more closely or expend more funds to keep the development process on track. A similar issue is whether contractors will undertake designs with substantial risk if there is a perception that such failures (perceived or real) will damage their credibility for subsequent competitions. Such a response could result in small satellite missions that resemble more traditional satellite programs. The many potential benefits of smaller missions cannot be realized if they are managed as traditional "large" missions, with the only difference being the size of the budget.

Hidden Programmatic Costs

There are costs associated with the management style discussed above that are frequently not considered and which comprise a special category of risk for small satellite missions.

The first such hidden cost involves the tendency of small, short development cycle missions, operating on low margins, to live off technical developments produced by much larger programs.[7] Small, low-margin missions cannot afford to explore and develop new technologies that have not yet been proven. They are, however, very adept at capitalizing on opportunities presented by other programs. For example, Microwave Monolithic Integrated Circuit technology, developed as an offshoot of various Department of Defense programs to improve millimeter-wave radar, is now being incorporated in the EOS Microwave Limb Scanner. Unless there is a continual research and development activity that can provide such new technologies, this source will eventually run dry, as the small missions typically extract these new developments but do not replenish them. Development of new technology requires time and funding. Larger programs have provided the development in the past because they had the margins and funds. Small programs can also provide the development, but only if the schedule and funding are made available. Cost, schedule, and technological innovation cannot all be optimized simultaneously.

In the past, much of the Earth remote sensing research and development was done under the auspices of NASA's Research and Analysis (R&A) program. This particular role of the R&A program has diminished considerably in the last decade.[8] Instead, the science community often focuses on what can be measured, rather than looking at the scientific questions that need to be answered with an eye to matching these with appropriate technology. Thus technology development often drives science rather than the other way around. New NASA programs, such as the Instrument Incubator concept, are attempting to bring together technologists and scientists in new ways so that technology is driven by an understanding of science issues and questions.

The second hidden cost is the human impacts of tightly integrated, small rapid development teams. While such a work structure can be exciting, it does exact a toll on workers through the demand for long hours, intense activity, and strong pressures for success (NRC, 1997a,b). Some people thrive in this "pressure cooker" environment, but many involved in small satellite missions report that they are unwilling to commit again to such projects for several years. This reluctance leads to turnover issues, as follow-on efforts could require entirely new teams—with the attendant learning curve costs and risks.

Scientific Risks

The small satellite approach carries with it several risks concerning scientific return:

1. Rapid development missions are often focused on "small" problems. Missions are not designed for long life and are sometimes viewed as "one shot" opportunities.
2. Missions employing small satellites are more likely to be developed as part of a program of technology demonstrations as opposed to a program in which the science return is paramount.
3. Small missions require a well-defined focus in order to keep them simple and costs low. This approach may not work well for scientific studies that require measurements of many processes.
4. Data processing and distribution may be related to relatively lower priority, thus making it difficult for nonproject scientists to gain access to the data. This problem could be exacerbated in the case of missions led by a principal investigator (PI) should research investigations become centered on an individual's personal scientific interests.
5. With more single-sensor missions, the proportion of funds spent on satellite hardware and launch costs will increase. Such funds might otherwise be spent on scientific research.

---

[7] For example, a recent report by the National Research Council states, "To succeed within their severe cost constraints, Explorer missions cannot afford instruments that require lengthy development or space qualification cycles. Therefore, the use of instruments and/or instrument subsystems that have been developed for previous missions is essential. The present funding cap on SMEX [Small Explorer] and MIDEX [Mid-size Explorer] could well prove too restrictive for building scientifically first-rate missions without such instrument heritage. Lessons learned from the space physics Explorers demonstrate the importance of instrument and spacecraft heritage in meeting science goals while remaining within cost and schedule limits" (NRC, 1997b, Finding 3).

[8] For a detailed examination of the role of R&A in NASA programs, see NRC (1998).

Measurements that are collected by small, focused missions will tend to be the vision of an individual or team rather than the vision of the broader Earth science community. While this approach has resulted in significant scientific progress in the past and will likely continue to do so, the increasing complexity and interdisciplinary nature of Earth science research requires larger and more broadly based scientific teams (NRC, 1997c). In fact, the scientific research community increasingly views remote sensing systems as shared resources rather than the province of remote sensing specialists.

Small satellite missions provide an increased level of programmatic flexibility throughout the funding process. Missions can be delayed or accelerated more easily because both the scientific constituency and the scope of the program are smaller. However, the relative ease with which small programs can be delayed can have scientific consequences that extend beyond a particular mission—for example, when an integrated observing system is being developed that depends on satellite clusters or constellations, or when anticipated scientific returns are driven by an assumed schedule of satellite missions. The danger is that critical measurements will not be made with sufficient quality, or for a sufficient period of time, or that necessary complementary data sets will not be available.

Management of Scientific Risks

Mitigation of the scientific risks associated with small satellite operations requires innovative program management. Indeed, an effective response to the challenges posed by inclusion of small satellites in mission plans can lead to a more effective mission as measured by the science return.

A calibration and validation strategy should be developed for every mission, as well as for the overall suite of observing systems. In an effort to lower costs and to accommodate the mission on a small platform and small launch vehicle, on-board calibration systems as well as extensive prelaunch characterization are sometimes omitted or cut short. However, many important processes in the Earth system have long characteristic time scales so that separation of natural variability from variability in the sensor system is a difficult process. This does not necessarily mean that every mission must have complicated (and costly) on-board calibration systems. Instead, a well-developed (and well-supported) plan to ensure scientifically based levels of calibration and validation must include all aspects of an observing strategy. This may include prelaunch characterization of the sensor, strategies to ensure dynamic continuity of the data stream as sensors change and evolve, and field programs to quantify data quality.

The science community should also assess science needs for precise calibration of individual sensors against the need for more intensive temporal and spatial sampling, especially in the context of constellations of small satellites. The total error of any data set will be a convolution of the quality of the individual measurements that go into the set and the sampling error of the data (e.g., unresolved processes because of inadequate sampling, biases because of cloudiness, etc.). Many Earth science data sets must be continued for the foreseeable future due to either operational requirements or the presence of long-term (interannual to decadal) fluctuations. A strategy is required to ensure that new technologies and new measurements can migrate to the more slowly evolving series of operational satellite systems, such as the National Polar-orbiting Operational Environmental Satellite System (NPOESS). This includes, for example, a means to ensure a balance between the use of innovative technologies and of flight-proven systems.

Data processing and distribution must be an integral part of any mission, large or small, although these mission elements may result in higher costs to accommodate the on-board storage and downlink capabilities necessary to acquire global data. For example, the Lewis mission design was severely restricted in its ability to collect data (roughly 200 scenes), which essentially precluded its use as a global observing platform. Significant processing capabilities (either on board or on the ground) may be needed to reduce the overall costs of a constellation approach, such as data assembly to provide global observation from narrow swath sensors or compressing of downlinked data streams.

Adoption of a constellation approach for an Earth observation mission will require a far different management and budgeting structure within NASA than that currently in place. While the operational agencies have experience in maintaining small constellations, NASA does not. An observing strategy based on a dozen or so small satellites that would need to be maintained for many years—such as Clouds and the Earth's Radiation Energy System type observations of Earth's radiation processes—will require different planning, development, and operating strategies than the typical NASA mission. There are many precedents, such as the Global Positioning System, the Defense Meteorological Satellite Program, and the Polar-orbiting Operational Environmental Satellite, that could provide important lessons if NASA decides to pursue such a course.

Hidden Scientific Costs

Increasing the number and decreasing the size of missions can have beneficial effects, but the increased burden on both contractors and government agencies should be recognized. For example, there will be a significant increase in time and effort for proposal preparation and evaluation. Smaller, shorter-time-duration missions imply more opportunities to propose new ideas and technologies. Since it is essential that the science community be involved in the evaluation process, this will mean more time being spent on reviews and panels. In addition, the overall observation strategy must be continually updated and evaluated against new opportunities to ensure that it remains relevant. Thus the benefit of having more missions is likely to be accompanied by an increased burden on the science community, as well as increased development and proposal costs.

With the trend toward flatter management structures in small satellite missions, PIs are being urged (if not required) to assume more and more responsibilities for the end-to-end system. In addition, with smaller profit margins, vendors of the various mission elements (launcher, satellite bus, etc.) are less willing to spend significant amounts of time in proposal development and support. Instead, basic information is provided, and it is up to the scientist to evaluate and assemble the components. The recent Earth System Science Pathfinder process resulted in over 85 initial proposals. Twelve proposals were selected for further development in a second round of evaluation. Finally, two proposals (and one backup) were selected. Although the two-stage proposal process is meant to reduce the proposal burden, this is often not the case. The amount of effort on the part of the PI to develop a credible initial phase proposal is significant. Given the low success rate in both phases, it is in the interest of the investigators to make the initial proposal as complete as possible.

The loss in scientific productivity from these increased burdens is difficult to quantify, but is likely not negligible. Any complete evaluation of the benefits of small satellite missions should attempt to account for this cost.

SUMMARY

Management innovations are needed to exploit the potential advantages offered by the small satellite approach. Maintenance of science quality must be foremost in implementing these changes, however. This in turn will require a science-driven (versus technology-driven) approach to small satellite missions, as well as development and implementation of strategies to maintain dynamic continuity between sensors on successive satellites. An overall strategy for Earth observation is needed to serve as the benchmark against which to evaluate new missions, especially if research and operational observing systems move toward a constellation approach.

Use of small satellites in either a smaller and faster or constellation manner will require management to rethink how it assesses and manages risk. Compared to large missions, management will need to tolerate higher levels of risk and develop a more flexible response to failure. The management of small satellite programs also needs to adopt a more streamlined and less hierarchical approach than is typical for larger missions. It is advantageous if interactions between contractor and government emphasize insight rather than oversight. Finally, smaller product development teams may lower costs, but this should be achieved by improving processes and increasing risk tolerance—not by increasing pressure on the team.

The Earth science community must adjust to these new approaches. Sampling strategies must be placed on an equal footing with the drive to improve sensor quality. The community must be willing to streamline its proposal development and review procedures. Operational observing systems such as NPOESS will play an increasingly important role in Earth system research along with the traditional NASA research missions, and the research community must evaluate the full spectrum of Earth remote sensing missions in the context of a coherent observing strategy.

REFERENCES

Fleeter, R. 1997. Mr. Murphy on small spacecraft and rocket reliability. Launchspace 2:14-16.
National Research Council (NRC), Space Studies Board. 1997a. Lessons Learned from the Clementine Mission. Washington, D.C.: National Academy Press.
National Research Council (NRC), Space Studies Board and Board on Atmospheric Sciences and Climate. 1997b. Scientific Assessment of NASA's SMEX-MIDEX Space Physics Mission Selections. Washington, D.C.: National Academy Press.
National Research Council (NRC), U.S. Global Change Research Committee. 1997c. Pathways. Washington, D.C.: National Academy Press.
National Research Council (NRC), Space Studies Board. 1998. Supporting Research and Data Analysis in NASA's Science Programs: Engines for Innovation and Synthesis. Washington, D.C.: National Academy Press.
Ruffner, K.C. 1995. CORONA: America's First Satellite Program. Washington, D.C.: Central Intelligence Agency History Staff. Available through the National Technical Information Service.
Shaiken, H., S. Lopez, and I. Mankita. 1996. Experienced Workers and New Ways of Organizing Work: A Case Study of Saturn and Chrysler Jefferson North. National Center for the Workplace Working Paper Series, No. 5. Berkeley, Calif.: Institute of Industrial Relations, University of California, Berkeley. Available online at <http://violet.lib.berkeley.edu/~iir/>.

8

Findings and Recommendations

Advances in technology and the response of the marketplace have led to smaller sensors, satellites, and launch vehicles capable of performing useful space missions at relatively low cost. The Committee on Earth Studies has examined the capabilities and potential roles of such small satellites in key National Aeronautics and Space Administration (NASA) and National Oceanic and Atmospheric Administration (NOAA) Earth observation programs. The committee's study focused in particular on the use of small satellites in the NASA Earth Observing System (EOS) program and the planned NOAA-Department of Defense National Polar-orbiting Operational Environmental Satellite System (NPOESS) program. This chapter reviews, in the broad context of an integrated study, the topics highlighted in the summary sections of the preceding chapters, and it presents key findings and recommendations.

MISSION COSTS

Earth observation mission life cycle costs are the sum of all those incurred for developing, fielding, operating, and maintaining all elements of the system; these include the sensor payload, satellite, launch vehicle, and ground segments (command, communications, control, and data processing). It is often difficult to determine the true costs of some of these mission elements because prior expenditures on technology development, system infrastructure, and even some mission components (e.g., available sensors) may not be accounted for. When assessing and comparing costs for alternative mission architectures, comparisons must be made on an equivalent basis and not be biased by unequal treatment of these "hidden" costs.

The lowest cost missions are achieved when the costs of all mission elements are minimized. For their space segments, the promise of low-cost small satellite missions is based on this approach—low-cost sensors accommodated on low-cost small satellite buses and launched on low-cost small launch vehicles. The committee found that small launch vehicles are available at substantially lower cost than the Delta or Atlas class used for mid-size or larger satellites, but at higher specific cost (cost per pound of satellite to orbit). Similarly, small satellite buses are available at substantially lower cost than the larger Delta or Atlas class satellites (depending on capability), but again at higher specific cost (cost per pound of satellite). Thus, low-cost sensors that can be accommodated on these smaller satellite buses and launch vehicles can be flown as low-cost small missions. Technology advances are helping reduce sensor size and cost (as long as complex deployables are not required). Simple missions with limited objectives will yield the lowest cost scenarios.

A different situation is extant in a trade between multiple small satellites versus a larger multisensor satellite to accommodate a given sensor payload. Here the higher specific costs for small satellites and small launch vehicles will generally result in a higher cost to field the system initially (but not necessarily to maintain it) than using a larger multisensor satellite and a matching launch vehicle. This is true irrespective of sensor size or cost.

MEETING MISSION GOALS: OPPORTUNITIES WITH SMALL SATELLITES

Much of the interest in small satellites stems from a desire to "do more with less" and an assumption that small satellite missions result in lower costs. A more pragmatic objective reflecting recent budget realities might be to "spend less and do as much as possible." Small satellites clearly provide a vehicle for accomplishing the latter. Here, the committee distinguishes between "small satellites" (100 to 500 kg), "small missions" (low cost), and larger (higher cost) missions that are performed with multiple small satellites. **Low-cost small satellites help enable low-cost small missions. This benefit is derived as much from the relative simplicity of many small missions (or the preexistence of mission elements) as from the size of the satellite.** Small missions generally consist of only one or a few sensors and may have less stringent requirements as measured by performance, calibration, or longevity. Less complex missions require shorter development time, which goes to the heart of lower costs. Because simpler missions may also be less capable, the science and operational needs must be carefully evaluated to ensure that they are adequately addressed.

When considering small satellites to perform a larger mission involving a number of sensors, a mission architecture trade-off study is required. Alternative architectures include accommodating all sensors on a single larger platform, on multiple small satellites, or on a mixed fleet. Trade-off criteria may include programmatic flexibility, preferred measurement sampling strategies, risk tolerance, system robustness, schedule, and—of course—life cycle cost. The lowest cost architecture for such missions is not evident a priori, but depends on mission-specific parameters.

One of the emerging benefits of small satellite missions is a reduction in the "time to science." Large, complicated missions often take many years to develop, during which time both scientific understanding (and hence requirements) and technology may evolve substantially. In addition, an increasingly cost-constrained fiscal environment makes large missions especially vulnerable to budget instabilities. When a large mission can be accomplished with multiple small satellites, this approach may lead to faster science return—but this is not guaranteed. The overall schedule and cost must be examined to determine if the need for multiple satellites and launches increases or reduces the time interval to establish full capability. The potential for obtaining some (perhaps the most important) data sooner can be a compelling driver.

Small satellites offer the potential for new mission architectures, such as clusters or constellations. Such architectures may permit development of new observing strategies that alter the relative balance between observation error (as quantified by parameters such as signal-to-noise ratio) and sampling error. Employing constellations of small satellites to acquire large amounts of observational data, albeit of perhaps lower quality,[1] may provide a more robust estimate of the overall statistics of the data field. This aspect of the scientific mission has not been examined in detail in this report and would have to be considered on a case-by-case basis.

OPERATIONAL AND RESEARCH EARTH OBSERVATIONS

Although there are differences between the operational measurement requirements of missions such as NPOESS and the science requirements of research-oriented missions such as NASA's EOS, there is clearly overlap as well. Moreover, many operational measurements are useful for research, especially for long-term climate studies. **The separation of instrument variability from the often subtle, long-term variations in climate-related processes requires careful calibration and validation of the sensor and its derived data products.** As sensors are replaced over time, it is essential to maintain "dynamic continuity" of the data product

---

[1] Data quality could suffer if the capabilities of the many required sensors were diminished as a result of cost constraints. The smaller accommodations provided by small satellite buses could also result in diminished sensor capabilities.

despite changes in sensor performance. As a result of the need to protect life and property, operational systems generally have little tolerance for temporal gaps. Research systems can generally tolerate longer gaps, especially in the area of climate research, as long as dynamic continuity of the data can be achieved through calibration. Cross-calibration of sequential sensors by ensuring temporal overlap of their satellite platforms is a preferred method of guaranteeing such dynamic continuity.

Except for studies of clouds, most of the research systems do not need strict simultaneity with co-boresighting of multiple sensors on a single platform. Rather, it is more important to ensure that a full suite of sensors is available to measure processes related to coupling of various components of the Earth system, such as air/sea fluxes, and that this suite is continued for a sufficient period of time. Thus, the emphasis in research is on contemporaneous data sets rather than strict simultaneity. For operational systems, strict simultaneity is also not generally required. Because the sensors are not all co-boresighted and because some have inherently different sampling strategies, even operational satellite platforms that carry multiple sensors mostly provide contemporaneous rather than simultaneous observations. Distributing appropriate groups of sensors (e.g., an atmospheric sounding suite) on a cluster of satellites flying in close formation would likely meet requirements for spatially coordinated measurements in operational weather systems.

The requirements for research missions evolve rapidly with advances in science and technology. Long development times often run counter to this emphasis on flying the latest in sensor design. Moreover, research missions emphasize the quality of the individual observation and thus constantly push the technology envelope in an attempt to obtain better quality data. Operational systems, on the other hand, tend to evolve more slowly, in part in response to budgets, which grow more slowly, and in part in response to the well-defined operational nature of the missions. For example, the data processing infrastructure of the operational user community often involves numerical models that may be expressly designed to assimilate satellite measurements collected at specific times with specific observing characteristics.

Small satellites offer new opportunities to partition a program's space architecture between small, focused missions and larger, more comprehensive missions. This flexibility is of particular importance to meet the differing needs of operational and research missions. For example, operational missions may use small satellites in a replacement strategy to ensure minimum gaps of critical data records, whereas research missions may use small satellites for maximum programmatic flexibility and to ensure minimum "time to science."

PAYLOADS

The potential to design smaller satellites for Earth observation missions is driven by advances in microelectronics and other technologies that facilitate the design of smaller and lighter sensors and spacecraft subsystems. However, there are fundamental laws of physics that in some cases restrict the degree of miniaturization that can be achieved while retaining sufficient performance to meet the observation requirements.

Sensor design and size are determined by a complex trade-off among spatial, spectral, and radiometric performance. All three of these performance measurements are interdependent, and, barring compensatory design changes, each measurement is improved at the expense of the others. Developing a design that balances all of these performance parameters while minimizing size, cost, and technical risk is the essence of sensor system engineering.

Within these constraints, technology improvements may alter the various design trade-offs and permit improved performance, lowered costs, and/or more compact sensors. **Reductions in sensor size, mass, and power can have substantial leverage on the entire space segment architecture and costs in that smaller sensors can be accommodated on smaller spacecraft and smaller spacecraft placed into orbit with smaller launch vehicles.**

The fundamental philosophy of sensor design should also reflect the architectural trades available with small satellites and focus more on specific observing tasks than on general applications. The more observation requirements that a particular sensor attempts to fulfill, the more complex the design. Such general-purpose sensors often must balance conflicting requirements, sometimes resulting in poorer performance than would be achieved by a design focused on a subset of the requirements. **Reduction in size and system complexity, as well as simpler** payload integration, can often be achieved by developing "task-specific" sensors. **When measurement objectives are limited, task-specific sensors on small satellites are a promising approach to low-cost, focused missions.**

SATELLITE BUSES

Technically capable small satellite buses suitable for Earth observation missions are now available. Further development efforts should continue to reduce power, weight, volume, and other aspects of these platforms as well as enhance their payload accommodation capabilities. Present science and operational payloads are sometimes too big, require too much power, or create too much vibration to be accommodated on these platforms, but the trend toward smaller sensors should expand their utility in the future. Depending on capability, small satellite buses to support 100 to 500 kg spacecraft are available at recurring costs in the range of $10 million to $30 million. **Low-cost "production" buses, developed primarily for telecommunications applications, generally must be tailored to meet the needs of Earth observation missions (for example, adding star trackers to provide more precise pointing); such customization can add substantially to costs.** To take best advantage of such buses, instrument development must focus more on building to a standard platform interface rather than customizing the interface (and the platform) to accommodate the sensor. Such an approach is the reverse of the way sensors are traditionally developed, where sensor requirements drive platform design, and care must be exercised not to compromise the science objectives.

LAUNCH VEHICLES

The launch vehicle must be matched to the mission if costs are to be minimized. Although the cost per kilogram of performance (maximum satellite payload mass to orbit) increases as the size of the launcher decreases, the total cost of the system will generally be lowest when launch vehicle performance is matched to satellite mass. Excess capacity represents wasted costs. **Present launch vehicle capabilities do not effectively span the range of potential payloads, and there are notable gaps in performance, especially at the low end of the range.** Fairing volume (which determines the stowed payload size) is also limited with small launch vehicles and sometimes determines the type and complexity of deployable systems such as antennas. More flexible launch systems are needed where volume constraints are less stringent. Launching multiple satellites with a single larger launch vehicle is a partial remedy to these problems but comes at the expense of programmatic flexibility —a potential feature of small satellite missions.

The present scarcity of reliable small launch vehicles is of great concern. In part, this lack of reliability is a result of the relative newness of these systems, and the situation should improve with time. Robust mission architectures can mitigate the problem in the interim. For example, where possible, parsing the payload among clusters or constellations of small satellites allows a mission to be fielded and/or maintained by small launch vehicles with only modest sensitivity to failure.

MISSION ARCHITECTURES

The relative merits of small, mid-size, and large platforms are a complicated function of the overall mission objectives, available budgets, and success criteria. These criteria are significantly different for research and operational missions. For example, operational systems are judged by performance, life cycle cost, and availability (the percentage of time the system can deliver timely data, often on demand). **Long gaps between observations or loss of a single critical sensor can result in mission failure.** With research missions, dynamic continuity and data quality as well as the flexibility to pursue new sensors and new science requirements are more important than just data availability. These differences may lead to different mission architectures.

Trade-offs between multisensor platforms versus multisatellite approaches arise whenever a mission requires multiple sensors on orbit simultaneously. Such trade-offs are analogous to the multimeasurement versus special-purpose sensor trades discussed under payloads. **The lowest initial cost to place a given suite of sensors on orbit** is likely with multisensor platforms. However, the architecture with the lowest life cycle cost to field and maintain an operational suite of sensors in orbit is mission specific, dependent on the design lives and reliabilities of the system elements (for example, sensors, spacecraft bus, launch vehicles, and ground segment). With fewer launches and satellites involved, multisensor platforms offer a higher probability of successfully placing all sensors into orbit. Although the impact of a launch vehicle or spacecraft bus failure on a multisensor platform is obviously severe, these events have relatively low probability as multisensor spacecraft are typically designed with redundant systems,[2] and candidate launch vehicles such as the Delta II have historically demonstrated high reliability. Managers of operational missions consequently typically choose the multisensor, common bus approach. **Small satellites flying in formation with the larger platforms can play a critical role in timely replenishment or augmentation strategies for these systems.**

For Earth science research missions, small satellites provide important flexibility to the overall program. They provide schedule flexibility when responding to advances in scientific understanding, changes in scientific priorities, or development of new technologies. New missions can be developed and launched without waiting for accommodation on more slowly evolving multisensor platforms. Small satellite clusters and constellations provide new sampling strategies that may more accurately resolve temporal and spatial variability of Earth system processes. New approaches to calibration may be possible as well. These and other characteristics of small satellites allow the mission developer to design a program that is more balanced between long-term monitoring observations and short-term measurements for research than would be possible if using only larger systems.

The design of an overall mission architecture, whether for operational or research needs, is a complex process and requires a complete risk-benefit assessment for each particular mission. A mixed fleet of small satellite and larger multisensor platforms may provide the best combination of flexibility and robustness, but the exact nature of this mix will depend on mission requirements.

SYSTEM MANAGEMENT

Small satellites present several opportunities for both research and operational Earth observing missions. First, as noted previously, small satellites can enable low-cost missions and support rapid deployment. Their relative simplicity allows them to take quick advantage of innovations in both science and technology, thus leading to a shorter time to science. Second, small missions may enable new sampling strategies based on clusters or constellations. Such strategies may significantly enhance the scientific quality of the resulting data set and represent a new approach in Earth remote sensing. Third, missions built on a mixed fleet of multisensor and single-sensor spacecraft enable new maintenance and replenishment strategies, which may be particularly important for operational missions such as NPOESS. To realize these benefits fully, the management structure for the missions must be properly aligned as well. **Small missions, which are generally less complex and can accept more risk, are best implemented with less oversight.**

Management overhead can easily slow system development or discourage innovation, thus inhibiting many of the advantages of the small satellite approach. This argues for a management structure that is streamlined and less hierarchical than that typically employed. Small, tightly integrated teams have an advantage in such an environment; care must be exercised in how the teams are managed to avoid "burnout." In this context, for example, reducing team size too much in an effort to lower overhead costs becomes counterproductive. Excessive stress on the team can also be reduced if management has an understanding of and plans for the greater risk of failure that is typical in the "faster, better, cheaper" approach. It is also advantageous if government involvement is participatory and is characterized more by insight than oversight.

---

[2] Larger spacecraft typically incorporate redundant components for key subsystems. The committee views redundancy more as a matter of money than of size. While it is true that *very* small spacecraft may simply not have room for redundant units, the spacecraft under discussion in this report mostly do. It is in the interest of minimizing cost that redundancy is reduced in developing "low-end"-type spacecraft buses. This could be done with larger satellites too—although it seldom is because the payloads are too valuable. High-end small satellites with 3- to 5-year mission lives are designed with redundancy in many components.

Implementing new mission architectures based on a mixed fleet requires changes in management as well. First, there must be appropriate mechanisms to ensure the design and maintenance of a coherent observing strategy. For example, in the EOS program, solicitations for new missions should be consistent with the overall science directions of the Earth Science Enterprise. Second, management must address the issues associated with maintaining dynamic continuity of long-term data sets where the specific sensors (and even measurement techniques) will change over time. For operational missions, this includes replenishment strategies to replace failed sensors. It also includes developing comprehensive plans for cross-sensor calibration, data validation, and prelaunch sensor characterization. Third, with the assistance of the science community, management should undertake quantitative evaluations of how data quality varies under different assumptions regarding sampling frequency and individual sensor quality. This evaluation should include an assessment of the impacts of data gaps as well as temporal and spatial resolution.

SUMMARY

There is no single best strategy that replaces larger satellites with smaller satellites. However, an overall mission architecture that effectively combines the elements of large, mid-size, and small satellites can now be developed for Earth observing programs. Mission architecture choices must be driven by the requirements of the eventual users of the data. **In planning for future missions, the committee recommends that NASA and NOAA consider the merits of small, medium, and larger satellites without prejudice, seeking the most appropriate system architecture based on mission requirements and success criteria.**

FINDINGS AND RECOMMENDATIONS

Appendixes

APPENDIXES

A

Statement of Task

ANALYSIS OF SMALL SATELLITE CAPABILITIES IN LIGHT OF SCIENCE REQUIREMENTS FOR CORE OBSERVATIONAL NEEDS IN EARTH STUDIES

Background In recent years, "faster, better, cheaper" has become the National Aeronautics and Space Administration's (NASA's) motto for future missions. New technologies have evolved in the Department of Defense (through the Strategic Defense Initiative and other defense activities) and to enable constellations of low-altitude communication satellites. In addition small, commercial launch vehicles are being developed. Proponents now assert that these advances offer new means for the conduct of Earth observations that offer lower costs and enhanced capabilities. They also assert that the technologies are applicable to the National Oceanic and Atmospheric Administration's (NOAA's) Geostationary Operational Environmental Satellites (GOES). Furthermore, pressures on the federal budget have produced calls for a complete revamping of the Mission to Planet Earth (MTPE) [now called Earth Science Enterprise—ESE] and its ground data processing element (both in the pre-2000 era that was targeted in earlier restructurings and more recently in a new reexamination of post-2000 plans) and in the annual cost of NOAA's operational satellites. In both instances, the issue of the small satellite capabilities and applicability arises.

NASA's MTPE and NOAA's Polar-orbiting Operational Environmental Satellites (POES) have evolved in an era in which launch vehicle costs were a major driver in overall mission costs. The POES are in the process of being merged with the Air Force's Defense Meteorological Satellite Program (DMSP). The DMSP satellites evolved in the same era and with the same underlying assumption of high launch costs. This was also an era in which the available launch vehicles did not pose a serious limit on payload and spacecraft mass or volume. As a result, both MTPE and POES/DMSP found the most economical overall design to be one in which several sensors were flown on the same spacecraft. This configuration also offered the advantage of multiple Earth sensors simultaneously observing some atmospheric and surface parameters. It is time to reassess whether earlier assumptions and technical approaches should now be changed due to new technology and the changed environment.

Plan The study will address the following questions:

1. What are the core observational needs for NASA and NOAA/DMSP?
2. Are there simultaneity of measurement requirements that necessitate that particular sensors are orbited on the same spacecraft or on two or more spacecraft that orbit in close proximity to one another (e.g., in a satellite constellation)?

STATEMENT OF TASK

3. What instrument technologies are currently planned for the conduct of these observations?
4. Have new payload technologies emerged that can reduce cost, payload mass, power requirements, other interface needs (views of cold space for radiative cooling, views of celestial objects for calibration, thermal dissipation, etc.), and/or data communications needs with respect to those that are planned?
5. Have new spacecraft technologies emerged that can enhance payload mass, power, and volume fractions or reduce overall system costs?
6. As a result of technological advances, are there new ways of aggregating or parsing sensor systems (including satellite constellations) that can offer advantages in the cost or effectiveness of Earth observations?
7. If such ways exist, what are their implications for the future of MTPE, POES/DMSP, and GOES in terms of both spacecraft and ground system configurations?

B

Effects of Technology on Sensor Size and Design

This appendix supplements the discussion in Chapter 3 on the physical limits of sensor sizing by examining the effects of technology with regard to building more compact sensors.

For the purposes of this discussion, consider the generic electro-optical sensor as a transducer that ingests photons and ejects a digital bit stream. Although there are many other classes of sensors that are of interest—such as active electro-optical lidars and active and passive microwave sensors—the fact that passive electro-optical sensors account for a major portion of the Earth Observing System (EOS) and National Polar-orbiting Observational Environmental Satellite System payload suites makes them a reasonable focus for this discussion.

Figure B.1 illustrates some of the key subsystems associated with the photons-to-bits transduction process; these are discussed in the following sections. The combination of the state of the technology and the first-order physics of these subsystems determines the potential for reducing sensors' overall size and mass.

SCANNING MECHANISMS

Many sensors employ a scanning mechanism in order to cover a wide field of view in object space while utilizing narrow field-of-view telescope optics. The current generation of Geostationary Operational Environmental Satellites, Polar-orbiting Operational Environmental Satellites, Defense Meteorological Satellites, and Land Remote Sensing Satellite (Landsat) instruments all exemplify this approach. The size and mass of their scanning mechanisms are determined geometrically by the telescope's aperture size and the desired object space field of view. Constraints or specifications regarding the geometric fidelity of the resulting scan pattern can also influence size. For example, some of the more compact scanning approaches can produce curved rather than rectilinear scanning patterns on the ground, and the tolerability of such distortions can greatly influence the size and mass of the scanning subsystem. Thus, there may be a trade-off between opto-mechanical size and on-board or ground-based signal processing complexity (to correct the distortion). In some instances, it may be desirable to exchange added complexity in image reconstruction and ground processing for a more compact space instrument, particularly if there is centralized ground processing of the instrument's data stream. Such system-level trade-offs must be carefully addressed on a case-by-case basis.[1] Other factors, such as the need to view calibration reference sources, can also have a significant effect on the size of scanning subsystems.

---

[1] For example, because digital orthorectification of imagery is now commonplace, there is little need for a sensor to collect a perfectly

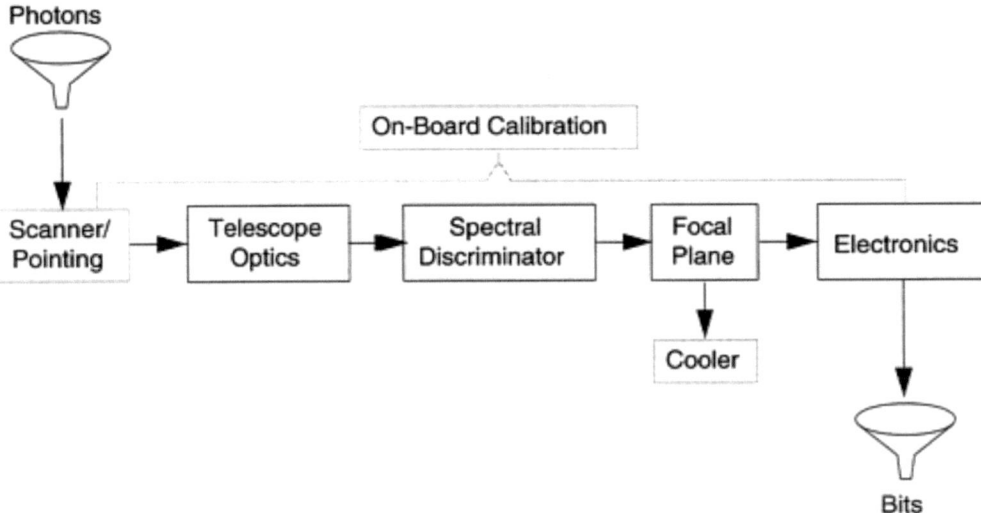

Figure B.1
Depiction of the generic electro-optical sensor.

Since the sizing of scanning mechanisms is driven principally by geometric considerations, technology plays a secondary role in determining size and mass. Technology is certainly a factor in developing lightweight scan mirrors, electromagnetic torquers, and position encoders, but these elements are already well developed and not subject to dramatic improvements. Lightweight beryllium scan mirrors are already in widespread use, for example.

In other cases, specialized performance requirements make it preferable to design sensors with a moving telescope assembly instead of a moving scan mirror. The SeaWiFS (Sea-Viewing Wide Field-of-View Sensor) is designed with a scanning telescope in order to minimize polarization sensitivity and, correspondingly, to maximize the radiometric fidelity of its ocean color measurements.

Still other sensor design approaches dispense with scanning mechanisms altogether by employing wide-field telescope forms. Such designs cannot accommodate more complex telescope optics and detector arrays, but this is often a reasonable trade-off given the current state of technology. Pushbroom designs, utilizing linear detector arrays and optical designs with a moderately wide field of view (~5°) in one dimension, have been in use for some time in sensors such as the French SPOT (Systeme Pour l'Observation de la Terre) satellites; the design remains attractive for high-resolution imaging systems with fields of view to 15°. "Fish-eye" lens designs can be used to provide wider field coverage, but the severe geometric distortion and limited spectral range of such designs usually make them more appropriate for nonimaging applications such as missile launch detection. Pushbroom designs are not suited to moderate-resolution wide-field systems such as MODIS (Moderate-Resolution Imaging Spectrometer) or AVHRR (Advanced Very High Resolution Radiometer), because the extended focal planes of such designs—coupled with relatively short focal lengths—lead to unworkable field angles and geometric distortion problems.

Moving the entire spacecraft with the telescope assembly remaining fixed relative to the body of the spacecraft is yet another scanning alternative, and is particularly suitable for single-sensor small satellites. Here, the

---

rectilinear image. In fact, this technology was developed for wide-angle aerial photography, which has substantial distortion over the field of view of the camera due to its wide field. This example illustrates the increasing potential to use advances in ground processing capability as a means of avoiding development of an otherwise more sophisticated and costly space-borne instrument.

sensor design is simplified because scanning and pointing mechanisms or mirrors are eliminated, with concomitant savings in instrument payload size, cost, and complexity. At the same time, additional demands are placed on the spacecraft's attitude control and determination system, with more stringent requirements for control of angular position and angular velocity. This also requires added spacecraft mass and power, since control moment gyros designed and sized to provide spacecraft agility are heavier and require more power than reaction wheels that are used solely for attitude control.

TELESCOPE OPTICS

The telescope opto-mechanical assembly represents the next major sensor subsystem, and technology certainly plays a role in the size/mass equation after the first-order sizing is established by the immutable laws of diffraction. Specifically, different optical forms and prescriptions vary widely in terms of their packaging efficiency. For a given aperture diameter, effective focal length, and level of performance, optical systems of different designs can differ tremendously in terms of number of optical elements, total mirror area (and hence mass), as well as different fabrication and alignment tolerances. Figure B.2 is a scale drawing that shows three different optical designs for a pushbroom imager with a 15° field of view: Each design has exactly the same first-order properties—i.e., the same aperture diameter and focal length. Yet the overall packaging volume and total mirror area are strikingly different for the three designs. These differences translate directly to sensor size and mass. The effects are further magnified because the layout dimensions and alignment tolerances strongly influence the size and mass of the associated optical metering structure—the structure that holds the optical elements in their required positions with the requisite precision. Most of the remote sensing instruments currently in development or production have exploited advances in optical design techniques in order to minimize size and mass.

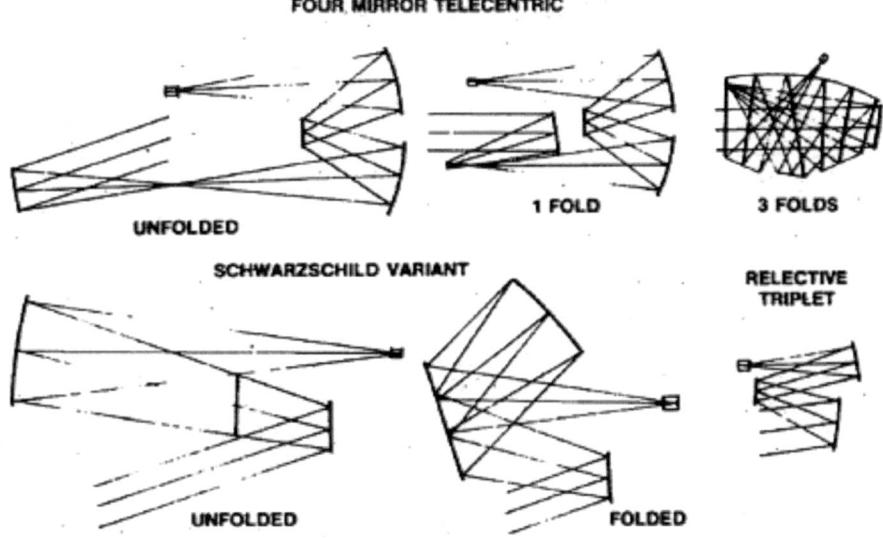

Figure B.2
This scale drawing of different telescope designs, all with the same aperture diameter and focal length, shows striking differences in packaging efficiency. Reflective triplet is most compact.

Besides clever optical design, advanced materials play an important role in offering future mass reductions. Lightweighting techniques are already widely used for both metallic and glass optical elements, and the use of high-stiffness-to-weight-ratio materials, such as graphite-epoxy composite, is the norm for optical metering structures. Further improvement in the materials arena should be of an incremental rather than revolutionary order, however. Silicon-carbide optics hold some promise, for example, but this will not lead to factor-of-two reductions in overall mass for typical EOS sensors.

As discussed above, the physics of diffraction is a key factor that constrains sensor design. However, there are techniques for mitigating the effects of diffraction that may offer some benefits for future high-resolution systems. Specifically, sparse aperture or partial aperture systems can provide diffraction limited performance that is comparable to that of filled aperture systems—albeit at the expense of photon-collecting area and, hence, signal-to-noise ratio. This may be a reasonable trade-off for systems in which resolution is more important than radiometric sensitivity.

As illustrated in Figure B.3, the partially filled aperture system on the right has essentially the same optical cutoff frequency as the full aperture system on the left, but the total mass of the mirrors would be dramatically smaller than a single large optic. Moreover, manufacturing several elements of modest size is less expensive and complex than manufacturing a single large element to exacting tolerances. In the particular case illustrated in the figure, the total mirror area of the sparse aperture system is just half of the conventional design, and the corresponding mass ratio would be even less than the area ratio because the mass has to increase on the order of the cube of the largest linear dimension in order to maintain the same relative stiffness. However, the partial (or multiple) aperture system would require added complexity in mounting and alignment of the additional optical

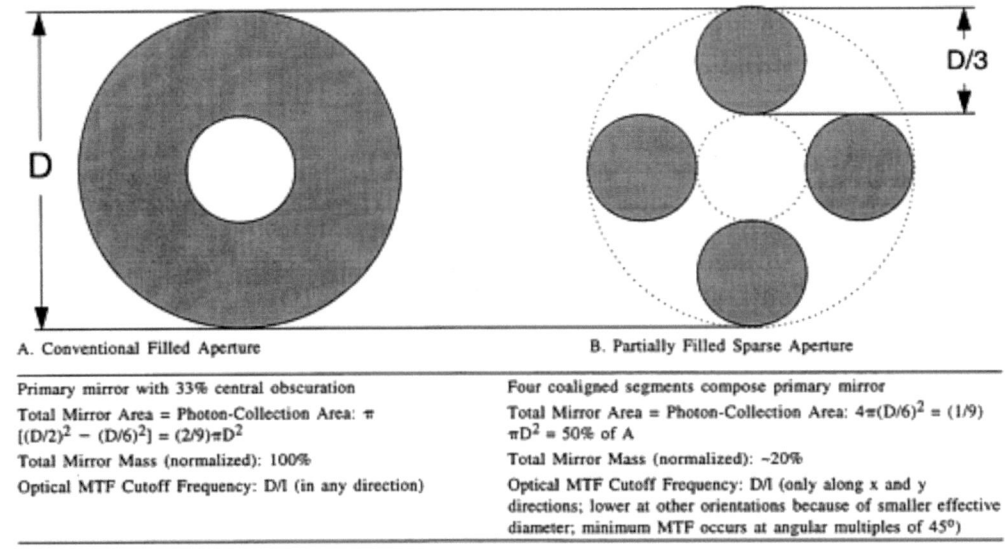

Figure B.3
Sparse aperture designs trade radiometric sensitivity and modulation transfer function (MTF) uniformity for lower mass.
Note: Mass scales approximately as the cube of the maximum linear dimension when designing to maintain approximately constant stiffness.

elements, and this would undermine some of the mass and cost savings. Further, there would be new technical issues in terms of maintaining the relative alignment of the segments, and there are some second order performance issues, such as modulation transfer function nonuniformity as a function of directional orientation.

Even more exotic techniques, such as optical synthetic apertures, can be considered methods for reducing at least the mass and size of telescope optics. Such techniques are clearly in their infancy: While synthetic apertures are practical at microwave frequencies, maintaining phase coherence (or an accurate phase history) is a formidable challenge at optical wavelengths.

SPECTRAL SEPARATION

Spectral separation is the next subsystem technology that is needed en route from photons to bits, and technological advances in this arena can certainly be enabling factors for size/mass reduction, particularly for hyperspectral instruments. For sensors having a relatively small number of spectral bands, multilayer dielectric interference filters remain the spectral separation method of choice. For this "multispectral" class of sensor, the technology trend has been toward multiple filters located near the focal plane.

As filter packaging techniques have improved, there have been some modest reductions in the volume of the aft optics and focal plane assemblies of these sensors, and some designers have explored direct deposition of spectral filters onto detector arrays. Although this is technically feasible, it is not in widespread use for radiometric instruments because it effectively cascades two complex processes, resulting in lower yields and consequently higher costs; the minor gains in packaging efficiency are generally not worth the added cost and risk. For example,

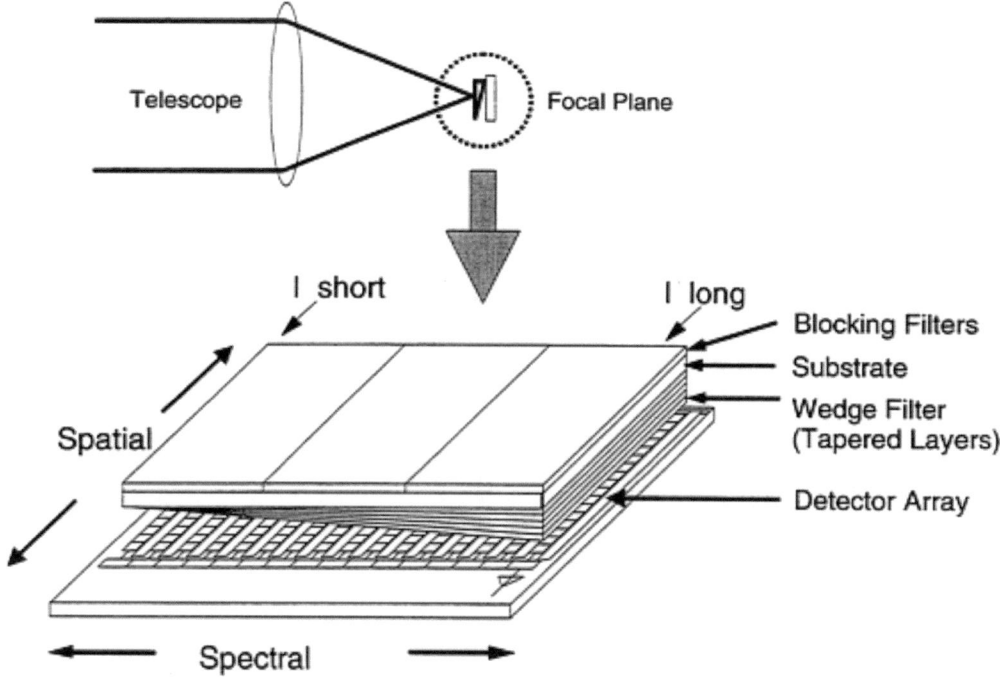

Figure B.4
Wedge spectrometer approach reduces size and mass of hyperspectral sensors.

if detector array processing is a 20-step process with each step having a yield of 0.98, the overall yield for detectors alone would be approximately 0.98^{20}, or 67 percent. If four separate filters each having 40 layers were subsequently deposited on the detector array, the overall yield of the detector assembly would be on the order of 2.6 percent (if the filter deposition yields were also 0.98 per layer).

Advances in filter technology are having very significant impacts on the design of hyperspectral instruments. Historically, hyperspectral spectrometers (i.e., instruments with hundreds of spectral bands) have been built with dispersive optical systems using either prisms or diffraction gratings. Such approaches work well, but they require large and complex reimaging optical systems. In contrast, spectrometers using tapered dielectric filters mounted close to the focal plane can dispense with the complex optics and make it possible to build very simple compact hyperspectral instruments, as illustrated in Figure B.4. While this "wedge spectrometer" design is attractive for many imaging and sounding applications, it is not a panacea: There are performance trade-offs regarding spectral simultaneity and sensitivity that may make conventional spectrometers more suitable for some applications.

FOCAL PLANE DETECTOR TECHNOLOGY

Detector technology can influence some basic sensor design trade-offs and lead to concomitant size and weight reductions. Increasing the number of detector elements can permit reductions in optical aperture size, until aperture size becomes governed by spatial resolution—and not by the optical collecting area required to achieve the requisite radiometric sensitivity. In essence, the availability of integrated high-density detector arrays allows designers to reduce the size of the optics to dimensions that are close to those set by the theoretical diffraction limit.

Technology now permits the design and production of integrated detector arrays that include first-level preamplification and multiplexing as part of the focal plane; this in turn permits simplification of the sensor's other electronic subsystems. For example, the focal plane arrays on the MODIS sensor utilize integral capacitive transimpedance amplifiers to integrate and amplify the detector signals; this eliminates the need for much larger off-focal-plane electronic modules that would otherwise be required to perform that function.

CRYOGENIC COOLERS

The cryogenic subsystems needed for cooling infrared detector arrays (and sometimes optical elements) are another key determinant of size and mass for sensors that extend spectral coverage into mid-to long-wave infrared wavelengths. There are relevant passive and active approaches to cooling. Passive radiative coolers are well developed and provide exceptional reliability. The size of these coolers is, however, governed by the first-order physics of Planck's radiation law. The required minimum area for the radiative surface is governed by the relationship $W = AesT^4$, where W is the radiated power, A is the area of the radiator, e is the effective emissivity of the radiator's surface, s is the Stefan-Boltzmann constant, and T is absolute temperature. This relationship can be expressed as $A = W/esT^4$. Thus, the area of the radiator needed to achieve a given cryogenic temperature depends directly upon the amount of cooling power required and upon the inverse fourth power of temperature. This is a highly nonlinear relationship; for example, it takes *four times* as much radiator area to reach 60 K as to reach 85 K. Consequently, application of this comparatively straightforward technology is limited to relatively modest levels of cooling power at temperatures of about 65 K or warmer. This is more than adequate for sensors such as MODIS, but passive cooling would be a challenge for a high-resolution pushbroom long-wave infrared imager, for example.

There are other considerations regarding the use of passive radiators. The radiator must be properly oriented and shielded to minimize thermal loading from the Earth, Sun, and spacecraft appendages in order to provide efficient cooling. There are also many technical subtleties in producing passive coolers in order to approach the theoretical minimum size set by Planck's radiation law. For example, control and minimization of parasitic heat loads is something of an art, but is essential to producing efficient radiative coolers.

Active cryogenic refrigerators offer an alternative to passive radiators. Active coolers offer greater capacity and provide additional freedom in packaging and locating the sensor on the spacecraft, since there are no preferred orientations or constraints on view factors. These benefits are provided at the expense of fairly high power consumption, added mass, and diminished reliability. Indeed, there is a crossover point in size/mass efficiency: Passive coolers tend to be the better choice for modest heat loads (<1W) at temperatures above ~80 K; active coolers become attractive for higher heat loads (>2W) at temperatures of 65 K or colder. There are application specific exceptions to these guidelines, and there is a gray region where the selection is not as clear cut, but the parameters cited above serve as a useful point of departure.

Active cryogenic refrigerators have been under development for many years with significant funding from the Department of Defense. This technology has progressed very well, with space-qualified, life-tested units available from British Aerospace, Hughes Aircraft, and TRW, among others.

ELECTRONICS TECHNOLOGY

Advances in electronics technology probably offer the most broadly and readily applicable avenue for further reductions in the size/mass/cost of space sensors. A substantial fraction of nearly every sensor design is devoted to the electronics that provide preamplification, filtering, analog-to-digital conversion, on-board signal processing, and high-speed multiplexing functions, as well as command, control, telemetry, and conditioned power. Although the size, mass, and power requirements of these electronics have been reduced at an astonishing rate over the past 20 years, such advances are often exploited to enhance functionality or reliability rather than size/mass reductions. And in fact, enhancing on-board processing capability can be a very logical choice from a total system cost standpoint. For example, in applications where a sensor provides a real-time data stream, it is almost certainly less expensive to perform radiometric responsivity correction on board the spacecraft instead of imposing a signal processing burden on each and every ground (or mobile) receiving station.

Even with added on-board functionality, the stunning improvements in electronic device technology hold the promise of significant reductions in size/mass/power requirements—with one caveat. The technological improvements are being driven largely by the commercial telecommunications and computer industries, and the supplier base for radiation-hardened, space-qualified components is, in fact, contracting. Thus, space systems have not been able to exploit fully the state of the art in electronics. Prudent investment in space qualification of commercial parts and direct investment in specific radiation-hardened devices will pay dividends in future size and mass reductions for space instruments.

Advances in electronic packaging technology also portend significant size/mass reductions for space-borne electronics. The migration to surface-mount packaging is already paying off, but next generation high-density multilayer integration techniques promise even more dramatic reductions in size. The essential idea is to move from the two-dimensional packaging approach embodied by circuit boards to three-dimensional packages that effectively stack or laminate multiple layers into a high-density package. This is more challenging than it might at first appear, because high-density packaging raises many technical issues such as thermal management and electromagnetic interference. Moreover, as packaging density increases, there are concerns about testability, repair and maintenance, and reliability—important factors because these "systems in a cube" become very high-value components, thus making the practicality and economics of testing and repair during the manufacturing process very important. Packaging technology is moving toward high-density three-dimensional modules, and the application of such technology to space instruments will have a major impact on overall sensor size and mass.

For example, electronics and their associated housings account for nearly 44 percent of MODIS's total mass. Therefore, as electronic device and packaging technology progress beyond the level embodied in the current MODIS design, retrofit of new electronics would be a highly cost-effective way to reduce the overall mass of the instrument while preserving the substantial nonrecurring design investment in MODIS's opto-mechanical subsystems, which are already governed more by physics than by technology.

ACTIVE SENSORS

In addition to passive electro-optical sensors, other classes of instruments, such as active optical systems and both active and passive microwave instruments, are of interest to Earth observation, meteorological, and planetary science missions. Much of the foregoing discussion regarding telescopes, pointing and scanning mechanisms, and electronics technology is equally applicable to both active and passive Earth observation sensors, but the design trade-offs and sizing constraints for active sensors are also heavily influenced by the technology of the active sources themselves—most specifically, lasers. In this regard, the overall "DC to photons" energy conversion efficiency is probably the most important factor in accommodating active Earth observing sensors on small satellites. This technology encompasses everything from high-efficiency solar arrays and DC to DC power converters to the efficiency of the laser devices themselves. These efficiencies vary widely depending on the wavelength of interest, but are typically in the range of a few percent for laser sources that are of interest to space science. The efficiency of the energy conversion process is the fundamental issue that will determine the future feasibility of small satellites to serve as platforms for sensors such as the Laser Atmospheric Wind Sounder.

Similar efficiency issues pertain to active microwave sensors, although there are additional degrees of design freedom wherein transmit power and antenna gain can be traded to achieve equivalent levels of effective isotropic radiated power. As antenna gain (directivity) is increased, transmitted power can be lowered, but the physical size of the antenna must grow proportionately to achieve the higher gain. There is thus a size/mass/power trade-off for active microwave systems. Pointing and scanning design issues comparable to those related to optical instruments are also involved.

C

U.S. Launch Vehicles for Small Satellites

This appendix provides the history and development status as of October 1998 for the candidate small satellite mission U.S. launch vehicles discussed in Chapter 5.

DELTA II

The Delta II is provided by Boeing Corporation (formerly McDonnell Douglas Aerospace). The company has provided various versions of the Delta vehicle since 1960 and has increased the payload capacity with each generation. The Delta II series became operational in 1989 and is capable of lifting 1,840 kg to geo-transfer orbit (GTO) in its heaviest configuration.

The Delta II launch vehicle consists of two primary stages with an option for a third. The last digit in the vehicle designation indicates which, if any, third stage is used; a version 7920, for instance, would indicate no third stage (Isakowitz, 1995, p. 230). The first stage main engine is a Rocketdyne RS-27, while the second stage uses the restartable Aerojet AJ10-118K engine; both engines use liquid fuel. The second digit in the vehicle designation indicates the number of solid rocket strap-ons (Hercules graphite epoxy motors—GEMs) the vehicle employs. Delta II 7925 uses nine strap-ons and a third stage, while the Delta II 7320 has three GEMs and no third stage. Models smaller than the 7920 are referred to as Delta II Lite vehicles.

The 7300 series is under contract to the National Aeronautics and Space Administration (NASA) under the agency's Medium Expendable Launch Vehicle (MELV, or Medlite) program. McDonnell Douglas (Boeing) is also looking into an opportunity to build a Delta Lite vehicle smaller than the 7300 series, consisting of two Castor 120 engines sequenced for stages 1 and 2, with an Aerojet AJ10-118K acting as the third stage. The first flight of a Delta Lite vehicle (7420-10) was on February 14, 1998, and carried four Globalstar satellites.

Currently, there are a variety of launch options for the Delta II. As mentioned previously, the Delta II 7925 is a true intermediate-class launch vehicle, capable of lifting 1,840 kg to GTO. However, the Medlite Delta II 7320 is capable of transporting 1,750 kg to a 700 km Sun-synchronous orbit. Even this smaller Delta II is half again as large as the next competitor, the Athena 2. A single Delta II 7320 flight could theoretically be used to boost up to three small Earth observing satellites simultaneously, contingent upon their relative size and orbital requirements.

The *International Space Industry Report* (1998b) estimates launch costs for the Delta II 7925 and 7320 at approximately $55 million and $35 million, respectively; NASA (1996, p. D-15) estimates the cost for launching on a Delta II 7320 is approximately $42 million for a single payload manifest under its MELV contract. Launch sites are Vandenburg Air Force Base and Cape Canaveral Air Force Station for Sun-synchronous and lower inclination orbits, respectively.

The Delta II is available with both a 9.5 and 10 ft fairing. The use of the 10 ft fairing reduces mass performance by around 50 kg for the three-stage vehicle and 120 kg for the two-stage launcher.

The Delta launcher has been a workhorse vehicle for both the civil and military space programs. It has the longest launch record of any family of vehicles in the American space program and has proven itself highly reliable. From 1989 through October 1998, the Delta II family had flown 67 times with 65 successes—a 97 percent success rate.

PEGASUS

The Pegasus launch vehicle is a commercially designed, all-solid propellant booster which is launched after being released from the belly of a Lockheed L-1011 aircraft. Designed in a cooperative effort between Orbital Sciences Corporation (OSC) and Hercules Aerospace, the launcher is now operated by OSC. The rocket is released from the belly of the aircraft when the plane reaches an altitude of 38,000 ft and a speed of Mach 0.79 (OSC, 1998, p. 2-1). The winged body launcher consists of three booster phases with an option for a fourth.

The use of the air drop technique grants OSC a level of launcher flexibility not enjoyed by ground-based launchers. Ground-based support is minimized, enabling OSC to launch basically from any site with an airstrip; the Pegasus is, in effect, a mobile launch system. In addition, the use of the aircraft allows the Pegasus to use the plane's velocity to gain a wider variety of orbital inclinations than can be obtained by a comparably sized vehicle launched from a similar ground site. In April 1997, the Pegasus successfully placed a Spanish research satellite in orbit, originating the L-1011 flight from the Spanish Canary Islands off the coast of Africa.

Since its maiden flight in 1990, OSC has marketed three variations on the Pegasus vehicle. The original Pegasus had three stages—respectively, an Orion 50S, an Orion 50, and an Orion 38. To increase performance and accuracy, OSC later added a fourth stage option, the Hydrazine Auxiliary Propulsion System (HAPS). To date, both flights with HAPS have resulted in less than nominal orbits, with the HAPS stage responsible for one of the anomalous results. OSC also made structural improvements to the first and second stages, enabling them to carry more propellant; it designated the improved vehicle the Pegasus XL, and the original Pegasus was subsequently phased out. The first Pegasus XL/HAPS was launched successfully in the latter half of 1997.

The Pegasus XL is capable of lifting a 225 kg payload to a Sun-synchronous orbit at 700 km altitude. NASA currently contracts with OSC to use the Pegasus vehicle under both the Ultralite Expendable Launch Vehicle Program and the Small Expendable Launch Vehicle Program. The former is for the launch of sub-150 kg payloads as secondary manifests on Pegasus flights; the latter is for a traditional dedicated Pegasus payload designation. Under these contracts, the cost of a Pegasus XL flight is $20 million for the dedicated launch and $8 million for the secondary manifest (NASA, 1996, p. D-15). OSC itself advertises a cost of $12 million to 14 million for an independently contracted launch.

Including all versions of the vehicle, the Pegasus has flown 24 times through October 1998. Of those flights, 19 achieved all launch objectives for a 79 percent total success rate.

TAURUS

The Taurus is also an OSC booster, first developed under a Defense Advanced Research Projects Agency (DARPA) contract for a demonstration launch of a "standard small launch vehicle" (NASA, 1996, p. D-15). The Taurus is a four-stage, all-solid rocket vehicle, which builds on the design of the Pegasus by adding an initial Castor 120 to the configuration and designating it "Stage 0" (minus the winged body that the Pegasus needs for air flight) (NASA, 1996, p. 257). The Taurus is designed to be launched from the ground as a mobile launching platform, capable of assembly and launch once on site in under a day; standard commercial service launches from established ranges.

The first Taurus debuted in 1994 with the successful launch of the Space Test Experiment Program M0/DARPASAT payload. The booster's second flight was in February 1998, when it carried three satellites into orbit.

Current OSC plans are to market a next generation version, the Taurus XL, which uses the same upper stages as the Pegasus XL. OSC intends to offer various versions of an upper stage, including the standard Orion 38 (now used on the Pegasus and Taurus) and the Star 37, which is larger and meant as a growth option. The Taurus XL/Orion 38 will launch 945 kg to Sun-synchronous orbit at 700 km. The Taurus XL/Star 37 will increase performance to the above orbit to around 1,160 kg. OSC plans to launch from Vandenburg Air Force Base for Sun-synchronous missions and from Cape Canaveral for lower inclination orbits. Reliability statistics in this case are not significant, as the Taurus XL has not yet flown, and the standard Taurus has flown only three times, albeit successfully.

OSC cites the cost of a dedicated manifest Taurus launch in the range of $18 million to $22 million. For its Medlite contract, NASA gives a cost of $30 million for the Taurus XL and $35 million for the XL version with an upper stage (Orion 38 or Star 37). The Taurus is available in both a 63 in. diameter fairing and a 92 in. one. The use of the larger fairing reduces mass performance by about 140 kg.

ATHENA (FORMERLY LOCKHEED MARTIN LAUNCH VEHICLE)

The Athena is an entirely commercial effort by Lockheed Martin to provide a family of launchers with incrementally increasing payload capacity. The Athena 1 is a two-stage, all-solid rocket vehicle, using a Castor 120 first stage and an Orbus 21D as the second stage, with an Orbital Adjustment Module carrying the Attitude Control System and avionics package. The next larger version, the Athena 2, made its maiden flight in January 1998 and adds an additional Castor 120 to the configuration. Farther in the future is the Athena 3, for which Lockheed Martin intends to add solid rocket Castor IVA-XL strap-ons to the Athena 2 design. The booster will launch from Vandenburg Air Force Base for polar orbits and from Cape Canaveral for lower inclination destinations. The Athena 1 has a capacity of 200 kg to a Sun-synchronous, 700 km circular orbit, while the Athena 2 can loft 700 kg and the Athena 3 with four strap-ons can loft 2,200 kg to the same orbit. Reliability statistics in this case are not significant, as the Athena 1 has flown only twice. The first flight, in August 1995, carried the GemStar 1 commercial payload but was a failure. The second, in August 1997, successfully placed NASA's Lewis space-craft into orbit. The Athena 2 made its first flight in January 1998, successfully launching the Lunar Prospector.

Recent estimates place the cost of an Athena 1 flight at $16 million, an Athena 2 at $22 million, and an Athena 3 at $30 million to commercial users (ISIR, 1998).

CONESTOGA

The Conestoga family of launch vehicles is assembled and operated by EER Systems Corporation. The Conestoga fleet is modular in design, with several variations intended to provide incrementally increasing payload capacity. In the Conestoga vehicle designation, the first digit indicates the type of core motor and the second the number of Castor IV strap-ons the version entails, which can be anywhere from two to six. This assemblage is topped by a mid- and upper-stage, designated by the third and fourth digits, respectively (Isakowitz, 1995, p. 220).

To date, only the Conestoga 1620 has been launched—in October 1995—and that flight ended in the destruction of the vehicle and its payload, the METEOR recoverable capsule. EER Systems has higher hopes for the smaller 1229, but without a payload backlog, the vehicle has an uncertain future. The company has designs to market larger versions of its rockets using a more capable core motor, but has not yet moved these into development.

The Conestoga 1229 has the capacity to loft 500 kg to a 185 km circular polar orbit. The larger 1620 can lift around 1,500 kg to the same orbit. However, EER Systems has no agreement in place to cover the use of a launch site capable of servicing these orbits. The company does have a contract to use Wallops Flight Facility for launches between 38° and 66° latitude, and plans to negotiate for the use of the Kodiak site in Alaska to service the higher inclination orbits should interest be shown.

EER Systems Corporation estimates a launch price between $18 million and $20 million for the 1620 (Isakowitz, 1995), and a cost of around $12 million for the smaller 1229 (Bille and Lishock, 1996, p. 9). Reliability estimates in this case are not statistically significant, since the Conestoga has had only one flight, and that launch resulted in the loss of the vehicle and its payload.

ECLIPSE, PACASTRO, KISTLER, AND EAGLE

There are a number of launch vehicles in the design and early development stages that merit consideration for their long-term effect on the small payload launch market and level of competition. If successful, any of the ventures discussed here could prove to be serious competition to the more established market players discussed above. Many of these proposals, especially those that utilize reusable or partially reusable designs would, in principle, have significant cost advantages over their more traditional competitors if they even come close to their stated objectives. Although data are more scarce on these proposed vehicles than on launchers currently in operation, a brief discussion of the stated objectives of each launcher or family of vehicles, in conjunction with their performance goals, will be a useful addition to the dialogue concerning the availability and cost of launchers for small satellites fulfilling Earth observation needs.

Eclipse Express and Astroliner

The Eclipse Express is proposed as a hybrid—part reusable, part expendable launch vehicle. The design calls for a modified F-106 drone to be towed by a Boeing 747 and released. The vehicle will then release at the apogee of its flight an expendable upper stage on which the payload is attached. The drone returns to Earth for the next flight. The initial towing capability demonstration is funded in part by a contract between Kelly Space and Technology and the U.S. Air Force. The launcher is limited in its payload capacity. For a 700 km circular orbit at 90° inclination, Kelly Space and Technology estimates a payload capacity of around 100 kg. However, Kelly's cost goal of $2 million per flight (Bille and Lishock, 1996, p. 9) would make it a competitive player despite its narrower payload capacity.

Beyond the Eclipse Express, Kelly Space has plans to market a more capable launch system, the Eclipse Astroliner, whose design is based upon the technological fundamentals proven in the Express program. Kelly Space estimates a payload capacity of approximately 1,590 kg to a 90° circular orbit at 463 km (SpaceDaily, 1998). No cost data were available as of this writing.

PacAstro

The AeroAstro Corporation is in the later development stages for a suborbital launcher, designated PA-X, that is jointly funded with the U.S. Department of Defense. The company intends to build on this vehicle to create three successively larger versions for orbital missions, the R2-10, the R2-150, and the R3-1000. The R2 vehicles are both two-stage, liquid-fueled expendable rockets. The R3 adds a third liquid-fueled stage. AeroAstro cites the proven reliability and greater safety of liquid-fueled rockets, as well as its simple stacked design, as key cost savers. The company currently holds contracts for 10 satellite launches, 3 of which are for KITComm (Australia) and the Swedish Space Corporation.

The R2-10, at a cost of $4 million, is capable only of launching "Bitsy-class" satellites. Its larger sibling, the R2-150, has a payload capacity of 250 kg to 370 km Sun-synchronous orbit. The R3-1000 projected payload capacity to a 705 km Sun-synchronous orbit is 450 kg. AeroAstro cites a cost of $6 million for the R2-150; cost figures for the R3-1000 are not yet available. While precise payload fairing information was not available, AeroAstro advertises the advantages of its wide and tall fairing in reducing the need for deployable structures.

Kistler

Kistler Aerospace has plans to design and market a two-stage, fully reusable launch system using entirely private funds. The initial system, called the K-1, will utilize three Aerojet/Russian NK-33 LOx/kerosene engines as the first stage and a single NK-33 as the second stage in a stacked design (Kistler, 1999a). Upon separation, the first stage will maneuver to a trajectory back to the launching site, touching down with a parachute-assisted landing. The second stage will follow a similar sequence upon separation from the payload. Kistler's plans call for an inaugural flight after 2000 (Kistler, 1999b).

The K-1 is designed to carry up to 900 kg to low Earth orbit (LEO). Longer term company plans call for a second generation vehicle, the K-2, to come into operation a few years after K-1 with an LEO capacity of around 2,700 kg. Kistler has expressed a desire to build an even larger version, with a 9,000 kg LEO capacity, but these plans are not yet well defined.

Eagle

E-Aerospace has plans to market a family of launchers based on the solid rocket motors used in the Peacekeeper missile. No further information was available as of this writing.

REFERENCES

Bille, M.A., and E. Lishock. Smallsat Launch Options: Choices and Challenges. 1996. Proceedings of 10th Annual American Institute of Aeronautics and Astronautics/Utah State University Conference on Small Satellites.

International Space Industry Report (ISIR). 1998a. July 6 issue. Available online at <http://www.launchspace.com/isir/home.html>.

———. 1998b. Nov. 9 issue. Available online at <http://www.launchspace.com/isir/home.html>.

Isakowitz, S.J., ed. 1995. International Reference Guide to Space Launch Systems, 2nd ed. Washington, D.C.: American Institute of Aeronautics and Astronautics.

Kistler Aerospace Corporation. 1999a. K-1 specifications and performance. Available online at <http://www.kistleraerospace.com/std/specs.html>.

———. 1999b. Kistler Aerospace development schedule. Available online at <http://www.kistleraerospace.com/std/schedule.html>.

National Aeronautics and Space Administration (NASA). 1996. Announcement of Opportunity: Earth System Science Pathfinder Missions. AO-96-MTPE-01. July 19.

Orbital Sciences Corporation (OSC). 1998. Commercial Pegasus user's guide. Available online at <http://www.orbital.com/Prods_n_Servs/Products/LaunchSystems/Pegasus/index.html>.

SpaceDaily. 1998. Kelly Space to demonstrate tow launch. Feb. 28 press release. Available online at <http://www.spacedaily.com>. Space News. 1999. 10(2):1.

D

Case Studies

The committee examined several cases in which small satellites and streamlined procurement and management approaches were employed to perform Earth observation missions. Candidate missions included the SeaWiFS (Sea-Viewing Wide Field-of-View Sensor), the TOMS-EP (Total Ozone Mapping Spectrometer Earth Probe), the Lewis and Clark missions under the SSTI (Small Spacecraft Technology Initiative), and the QuikSCAT (Quick Scatterometer) mission. For some of these missions, the data available to the committee were limited due to proprietary or other considerations, and only brief synopses are presented for the insights they provide. More complete case studies are included where data were available.

TOMS-EP

Program Objectives and Context

In support of global change research, the National Aeronautics and Space Administration (NASA) has been monitoring changes in the Earth's ozone layer with a series of ozone mapping spectrometer instruments on various spacecraft. The mission objective for the TOMS-EP satellite was to fill a gap and ensure continuity of data between similar instruments on the Russian METEOR satellite and Japanese ADEOS (Advanced Earth Observing Satellite). Continuity of coverage allows for better correlation of the thickness of the ozone layer with events on the Earth and Sun.

Program Alternatives

The TOMS-EP program was originally planned as an in-house Goddard Space Flight Center (GSFC) project; after some deliberation, the center decided that a competitive procurement was a better approach to meet the mission goals. The procurement was for a dedicated small satellite to accommodate a previously developed ozone mapping spectrometer from Perkin Elmer (now Orbital Sciences Corporation [OSC]). The contract was let in September 1991. The payload was procured by GSFC and supplied as government furnished equipment. GSFC selected the Pegasus XL to launch the satellite from the western range at Vandenburg Air Force Base into a 955 km Sun-synchronous orbit during the summer of 1994.

TRW was selected as the TOMS-EP contractor. Its proposed design was based on its Space Test Experiment Program (STEP) bus, upgraded to meet the reliability and life goals (0.90 at 3 years) of this critical mission. The upgrade was achieved through design improvements, parts upgrades, addition of redundancy, and a more robust quality assurance program.

The need to upgrade the bus led to an interesting programmatic trade-off. TRW's STEP spacecraft were being produced in facilities operating efficiently with streamlined processes appropriate to low-cost technology demonstration missions where higher levels of programmatic risk are acceptable. In contrast, processes at TRW's primary Space Park facilities were appropriate to the more demanding requirements of the high-reliability, performance-critical spacecraft produced at that site. Because its design was based on the STEP bus, consideration was given to developing TOMS-EP within the STEP facilities, using processes and controls modified to meet the more demanding TOMS-EP requirements. This plan was rejected, and the Space Park facility was selected for development, primarily to avoid technical and schedule risks to the program that might accrue from imbedding a high-reliability development program within a more informal culture.

Selected Approach

The key technical issues in design approach involved structure, solar array orientation/articulation, battery size, distributed versus centralized architecture for the data system, and the design complexity of the spacecraft safing mode. The finished product resulted in an aluminum structure with fixed arrays, a centralized data system, and a safing system that relied heavily on ground operations for recovery.

Recurring design trade-offs for TOMS-EP were the degree of design flexibility and the type of design margins to incorporate. Flexibility and large margins reduce risk and increase the potential for reuse of the bus design on future missions, but at increased cost for TOMS-EP. Because cost was an important issue on TOMS, most trades were decided in favor of limiting flexibility and margins to that needed to ensure the mission. Nonetheless, the TOMS-EP bus provided the heritage for several later spacecraft buses, including the Republic of China's ROCSAT-1, the Republic of Korea's KOMPSAT, and the SSTI Lewis satellites.

Status and Evaluation

The original plan was to launch TOMS-EP during the summer of 1994. Problems with the Pegasus XL launch vehicle delayed the launch to July 1996. At this time, the launch of the multisensor ADEOS with another TOMS instrument was imminent (it was launched August 17, 1996). The flexibility inherent in dedicated small satellite missions gave NASA the opportunity to reoptimize TOMS-EP to take better advantage of its concurrence with ADEOS. Thus, the TOMS-EP orbit was lowered from 955 km to 500 km where it would provide higher resolution data and augment the ADEOS science data return.

TOMS-EP was successfully launched and deployed on July 2, 1996. By mid-August, the spacecraft had gone through its integral propulsion system firings to get into the correct orbit, instruments were turned on, and TOMS became fully operational with real-time data available to the science community. TOMS-EP continues to be operational as of this writing.

The ADEOS spacecraft failed in orbit on June 29, 1997; lost with it were the data from the TOMS and other instruments it carried. Because the TOMS-EP spacecraft carries on-board propulsion, NASA could raise its orbit closer to that of ADEOS, both to increase coverage of the instrument and to reduce drag (and extend orbit life). The boost maneuver was performed in December 1997 and TOMS-EP was raised from a 500 km to a 750 km orbit. This will extend the mission's orbit life beyond the 2-year requirement and 3-year goal to as long as 5 years.

Lessons Learned

The TOMS-EP project embraced a low-cost, small satellite approach to flying a TOMS instrument over the 1994–1997 time frame as a potential gap filler to ensure continuity of ozone measurements between instruments on the Russian METEOR and Japanese ADEOS satellites. Key programmatic decisions were made, and the program plan was developed, to meet the performance and cost objectives on the desired schedule. TOMS-EP is a successfully operating system yielding valuable data. However, the program experienced both launch delays and cost growth. Some of the lessons learned or relearned were as follows:

- Early agreement and a freeze on requirements are essential for cost and schedule control. Both customer and in-house requirements "creep" are to be avoided.
- TRW used its Eagle Test Bed, an engineering model "spacecraft on a table," to expedite the project schedule through early verification of subsystem interfaces, validation of flight software, etc. This test bed proved very effective in reducing the development schedule and cost. It would have been even more effective had it been built with full redundancy rather than with single-string subsystems. Issues of redundancy management had to await the actual protoflight hardware for resolution.
- Fixed price commercial launch vehicles require discipline—in terms of early resolution and documentation of all requirements, interfaces, and other issues—in dealing with launch facilities, integration procedures, safety, and team working relationships. This ensures that the terms of the fixed price contract are clearly understood. A launch vehicle liaison—a designee on both sides of the interface responsible for working on and resolving all issues—should be identified early in the program. This liaison should document all results in a formal launch vehicle Interface Control Document; this too should be accomplished as early as possible.
- Program schedule and cost depend on all program elements. For TOMS-EP, the instrument and spacecraft were available on time to support the mission, but the launch vehicle was not. Significant expenditures made to deliver the satellite on time were not effective in meeting the mission schedule. Additional costs were then incurred to store the satellite and reconfigure the team to support the launch 2 years later.
- Risk must be carefully assessed for all program elements when defining the system, particularly for schedule-critical missions. For greatest cost-effectiveness, risk should be continuously assessed, progress monitored, and plans adjusted to keep the total program in balance.
- The TOMS-EP program clearly demonstrates the programmatic flexibility inherent in dedicated small satellite missions. The planned orbit was lowered to maximize synergy with the ADEOS mission; later, upon ADEOS failure, the actual orbit was raised to reoptimize the mission and extend its life.

SeaWIFS

Program Objectives and Context

The first research-quality space-based ocean color remote sensing instrument was the Coastal Zone Color Scanner (CZCS), which was launched on board Nimbus-7 in 1978. CZCS imagery revealed the presence of intense mesoscale variability in the spatial patterns of phytoplankton biomass; with the emergence of research on global biogeochemistry, it was realized that satellite ocean color data were an essential component of an ocean observing system. Planning for a replacement ocean color sensor soon began in the early 1980s as the CZCS was designed to operate for only 2 years.

The specifications for an ocean color sensor were well understood, based on analyses of CZCS data as well as in situ observations. By the mid-1980s, the International Geosphere Biosphere Program had assumed sponsorship of the International Joint Global Ocean Flux Study (JGOFS) which was focused on the role of the ocean in global biogeochemistry. One of the JGOFS objectives was the global study of ocean primary productivity and its variability, and satellite ocean color measurements were deemed to be an essential element of the program. This increased scientific support for a satellite ocean color mission culminated in an agreement whereby NASA agreed to purchase from OSC high-quality global ocean color data to be acquired by the SeaWiFS instrument. SeaWiFS would fly on OSC's SeaStar satellite. OSC would be responsible for SeaWiFS development as well as SeaStar development, launch, and operations and the provision of ocean color data to NASA. The SeaWiFS sensor would be built by Hughes (now Raytheon) Santa Barbara Research Center (SBRC).

Program Alternatives

A broad range of programmatic alternatives were explored for maintaining continuity of ocean color data. At one end of the spectrum, there were many attempts to secure sponsorship for a traditional government procurement wherein NASA, the National Oceanic and Atmospheric Administration (NOAA), the Department of Defense, or some combination of agencies would fully fund the development of an ocean color sensing satellite. At the other extreme, market studies and business plans were pursued to explore the feasibility of a fully commercial enterprise that would raise its own capital investment and subsequently sell ocean color data to all users. Between these extremes, many hybrid approaches were discussed and evaluated, including various government-industry partnerships such as shared investments, advance data purchases, guaranteed data purchases in the future, and "anchor-tenant" arrangements. The purpose of all these exploratory exercises was to find the most economical and expeditious avenue for providing the ocean color data stream. Because the budget authority for a traditional development program was not available, there was an avid search for alternative approaches that would achieve the same result for a smaller taxpayer investment. A stand-alone ocean color mission was expected to cost on the order of $70 million to $100 million, so the search was on to provide the same data stream to the science community for half the cost or less.

Timeliness was another critical dimension. Major international oceanographic studies were planned for 1993–1994 (such as JGOFS), and the availability of worldwide ocean color data from space would significantly enhance theses studies.

In this context, a number of players from government, industry, an academia were engaged to varying degrees in the search for a workable solution to the ocean color mission (see Table D.1).

Fully commercial approaches proved to be unworkable because an attractive business case could not be developed. The viability of a commercial offering depends upon data sales producing a revenue stream that permits recovery of the capital investment and provides a reasonable return for the investors, and such a case could not be constructed because the future revenue stream from the sale of ocean color data was unproven and speculative.

As various business and technical approaches were explored during the 1986–89 period, the most promising and economical scheme appeared to be a "piggyback" ride for an ocean color sensor on some polar-orbiting spacecraft under development for another mission. The essence of this idea was to fly the ocean color instrument for the incremental cost of accommodation on a spacecraft rather than bear the stand-alone cost of a dedicated spacecraft and launch vehicle. Candidate spacecraft included:

- NOAA/Earth Observation Satellite Company (EOSAT) Land Remote Sensing Satellite (Landsat) 6,

Table D.1 Organizations Interested in Ocean Color Data

Government	Industry	Academia
U.S.: NASA, NOAA, Navy, National Science Foundation	OSC	Joint Oceanographic Institute
Canada: Canadian Space Agency	Hughes Aircraft	American Geophysical Union
Australia	Telesat Canada	Scripps Institution of Oceanography,
Former Soviet Union	Nisho Iwai	University of California-San Diego
Japan	MacDonald-Detweiler	Woods Hole Oceanographic Institution
	Earth Observation Satellite Company	University of Miami,
	Fairchild Space	University of Southern Florida,
		University of Oregon,
		University of Washington,
		University of Rhode Island, etc.

- the Canadian Space Agency's Radarsat, and
- NOAA's Advanced Television Infrared Observation Satellite.

The first of these alternatives appeared to be the most attractive, and a deal was nearly signed between NASA and NOAA/EOSAT to fly a SeaWiFS on Landsat 6, but the arrangement ultimately fell through because of disagreements about liability for program delays amid concerns that the development time for the payload and modifications required to accommodate SeaWiFS would delay the Landsat 6 launch. For reference, the cost of this piggyback approach was estimated to be between $25 million and $35 million.

Selected Approach: Procurement of a Future Data Stream

Following the collapse of the Landsat 6 piggyback approach, NASA, Hughes, and OSC began exploring a stand-alone "lightsat" solution for the ocean color mission—one that would be as commercially oriented as possible. This process ultimately led to a competitive procurement for an ocean color data stream, wherein NASA would pay in advance for ocean color data to be provided in the future to the scientific community.

This was a significant departure from prior practice. Instead of purchasing hardware and software, NASA was buying only the promise of a future data stream (of specified quality). The value of the procurement was $43.5 million; most industry observers believed that this was indeed a bargain, because the actual cost of developing and fielding the system would probably exceed that figure. The balance of these costs would be recouped by the contractor through commercial sale of the data to the fishing, shipping, and offshore exploration industries, among others. This dual-use approach, where one sensor and spacecraft would serve both scientific and commercial customers, was made possible by tailoring the data distribution policy. Ocean color data are highly perishable, and their commercial utility declines rapidly with time because of the highly changeable nature of the ocean. On the other hand, most of the scientific research entailed is conducted by analyzing the data retrospectively, so it was easy to develop a scheme whereby commercial users would receive encrypted data in real time (for a fee), while scientific users would have unlimited access to the data after 48 hours.

After nearly 5 years of debate and evaluation of alternatives, a firm fixed price contract was let to OSC in March 1991. OSC subsequently contracted with Hughes SBRC to build the payload instrument, SeaWiFS. The objective was to have the spacecraft on orbit and delivering ocean color data by August 1993 in time to support JGOFS.

Status and Evaluation

Following award of the contract, OSC and SBRC proceeded with the design and development of the spacecraft, payload, and associated ground system. The schedule was extremely aggressive, but project participants expected that application of commercial practices and streamlined oversight would make it possible to meet the delivery date. The critical payload had a 2-year delivery goal—nearly unprecedented for the development of a new space-qualified electro-optical sensor.

The payload contract was let in May 1991; 24 months later, the completed SeaWiFS instrument was ready for delivery. The sensor met the performance specifications that had been established at the outset of the contract. However, the test data revealed that the sensor had significantly higher off-axis response which would degrade the scientific utility of the data. This posed a challenging technical and business dilemma. Although a specification-compliant sensor had been produced, the spec had failed to capture an important aspect of performance desired by the scientific community. If this had been a typical cost-type development program, the cost of subsequent work to further improve the instrument's performance would have been borne by the government if such an improvement was desired. Since this was a firm fixed-price commercial contract, and the letter of the payload specification had been satisfied, the payload contractor had no further obligation. However, after several discussions with NASA and the scientific community, OSC and Hughes decided to proceed at their own expense with sensor modifications to improve the instrument's performance. SBRC proceeded to incorporate sensor modifications that substantially improved the performance of the instrument to the satisfaction of the science community. The modified and retested instrument was delivered to OSC for spacecraft integration in December 1993.

Subsequent problems with spacecraft development at OSC led to protracted schedule delays, and the sensor remained in storage awaiting completion of the spacecraft. More recently, problems with the Pegasus XL launch vehicle have caused further delays. The SeaStar spacecraft remained on the ground nearly 6 years after contract award and more than 3 years after delivery of the payload. On August 1, 1997, the SeaStar (now known as OrbImage-2) spacecraft was successfully launched on board the Pegasus XL. Shortly thereafter, SeaWiFS began to deliver data and has now produced a nearly continuous time series of high-quality ocean color measurements since October 1997.

Lessons Learned

A number of salient points emerge in reviewing the SeaWiFS development initiative:

- The program reflected one of the classic flaws of many programs: A tremendous amount of time (nearly 5 years) was consumed in discussion, debate, and evaluation of program alternatives, leading to a highly compressed 2.5 year schedule for actual program execution—a schedule that ultimately proved unrealistic.
- There is substantial risk is proceeding with unproven, untested designs for space-borne systems and particularly for launch vehicles.
- The demonstrated ability of a contractor to build and fly space systems is the best assurance of success.
- Cooperative international research programs (such as JGOFS) cannot depend (or be predicated) upon the timely availability of data from developmental systems.
- It is difficult, if not impossible, to craft a specification that adequately constrains and establishes suitability of a product for its intended use. As a result, there is no substitute for close cooperation between data users and system builders. This ultimately worked well during the sensor development process on SeaWiFS, but there were some difficulties along the way in reconciling commercial and scientific imperatives.
- It is difficult to develop business arrangements, data policies, and operational protocols that concurrently satisfy government, industry, and academia. Hybrid structures such as government-industry partnerships and advance data purchases become problematic when one or more of the parties cannot deliver as promised.
- It is possible to develop space instruments on an accelerated schedule by adopting efficient design and oversight practices. Two years from contract start to flight hardware (31 months after performance-enhancing modifications) is much shorter than the traditional 4- to 5-year time line for payload instrument development programs.

SSTI (LEWIS AND CLARK)

SSTI was developed by NASA's Office of Space Access and Technology to advance the state of technology and reduce the costs associated in the design, integration, launch, and operation of small satellites. In July 1994, NASA awarded contracts to both TRW and CTA Space Systems to design and launch small Earth observing satellites named Lewis and Clark, respectively. Both contracts called for substantial new technology infusion into both payload and spacecraft bus, and for delivery of the satellites to launch within 24 months of contract start.

Both missions were unsuccessful. In the case of Lewis, the satellite development was completed within the allotted 24-month period; and, after a 1-year delay before its Athena 1 launch vehicle was deemed flight ready (see Appendix C), was successfully placed into its initial orbit in August 1997; the satellite, however, was subsequently lost. The Clark mission suffered excessive schedule delays and projected cost growth, ultimately leading to termination of the contract. Hopefully, much will be learned from their respective failures.

Lewis

A retrospective on the Lewis mission was provided by NASA in a synopsis of the report by the NASA-commissioned Lewis Spacecraft Mission Failure Investigation Board (1998). The report indicates that NASA's

Earth orbiting Lewis spacecraft failed due to a combination of a technically flawed attitude control system design and inadequate monitoring of the spacecraft during its crucial early operations phase.

Lewis was launched on August 23, 1997, with the goal of demonstrating advanced science instruments and spacecraft technologies for measuring changes in Earth's land surfaces. The spacecraft entered a flat spin in orbit that resulted in a loss of solar power and a fatal battery discharge. Contact with the spacecraft was lost on August 26; it then reentered the atmosphere and was destroyed on September 28.

The design of the Lewis attitude control system was adapted by TRW from its design for the system on the TOMS-EP spacecraft. The failure board found that this adaptation was done without sufficient consideration for applying the system's design to a different primary spacecraft spin-axis orientation on Lewis. As a result, minor rotational perturbations, possibly due to small imbalances in the forces produced by the spacecraft's attitude control thrusters, caused the Lewis spacecraft to enter a spin. This situation eventually overloaded the spacecraft's control system while it was in a safehold mode. Prelaunch simulation and testing of the spacecraft's safehold modes also were flawed because they failed to analyze this possibility.

The combination of these errors with the subsequent assumption that a small crew could monitor and operate Lewis with the aid of an autonomous safehold mode, even during the initial operations period, was the primary cause of mission failure, according to the failure board's report.

The failure board also assessed the role of the "faster, better, cheaper" project management approach in the Lewis program. "The Lewis mission was a bold attempt by NASA to jumpstart the application of the faster, better, and cheaper philosophy of doing its business," said Christine Anderson, chair of the failure board and director of space vehicles for the U.S. Air Force Research Laboratory at Kirtland Air Force Base, New Mexico (Isbell, 1998). "I do not think that this concept is flawed. What was flawed in the Lewis program, beyond some engineering assumptions, was the lack of clear understanding between NASA and TRW about how to apply this philosophy effectively. This includes developing an appropriate balance between the three elements of this philosophy, the need for well-defined, well-understood, and consistent roles for government and industry partners, and regular communication between all parts of the team."

Dr. Ghassem Asrar, NASA associate administrator for Earth Science, said, "The Lewis failure offers us some valuable lessons in program management and in our approach to technical insight. Lewis was an extreme example of allowing the contractor to have engineering autonomy. In the end, however, NASA has the responsibility to assure that the project objectives are met, and our assurance process was ineffective in this case. NASA's Office of the Chief Engineer is developing general lessons learned from this project and other faster, better, cheaper efforts, and we intend to apply them vigorously to all of our future missions, including the second generation of spacecraft in the Earth Observing System."

Clark

NASA issued a press release on February 25, 1998, announcing the termination of the Clark Earth science mission. The mission was terminated after an investment of some $55 million "due to mission costs, launch schedule delays, and concerns over the on-orbit capabilities the mission might provide" (Steitz, 1998). NASA is retaining Clark's launch vehicle services (an Athena 1 expendable launch vehicle). At the time of termination, NASA's contract for Clark was with OSC, which had earlier acquired CTA Space Systems to whom the contract had originally been awarded.

QuikSCAT

NASA's QuikSCAT mission was developed in response to the loss of the NASA Scatterometer (NSCAT) upon the failure of the ADEOS spacecraft. NSCAT data had proven its value in weather forecasting, and a replacement source was desired as quickly as possible. NASA was able to configure a replacement mission with a planned launch late in 1998—only 15 months from the start of the effort.

This very aggressive schedule was possible as a result of special circumstances. The Jet Propulsion Laboratory (JPL) would provide a scatterometer based on the future SeaWinds mission design by using available scatterometer hardware (with subsequent replacement for SeaWinds). Similarly, Ball Aerospace committed to provide a spacecraft bus in only 11 months using existing hardware which had been planned for an Earthwatch satellite (Quick bird) that had been put on hold following the loss of Ball's Earlybird. JPL used GSFC's Rapid Spacecraft Acquisition initiative (Chapter 4) to procure the RS2000 spacecraft bus quickly from Ball.

QuikSCAT was successfully launched in June 1999.

REFERENCES

Isbell, D. 1998. Lewis Spacecraft Failure Board report released. Release 98-109. Washington, D.C.: National Aeronautics and Space Administration.

Lewis Spacecraft Mission Failure Investigation Board. 1998. Available online at <http://arioch.gsfc.nasa.gov/300/html/lewis_document.pdf>.

Steitz, D. 1998. NASA terminates Clark Earth Science Mission. Release 98-35. Washington, D.C.: National Aeronautics and Space Administration.

E
Acronyms and Abbreviations

ADEOS	Advanced Earth Observing Satellite
AIRS	Atmospheric Infrared Sounder
ALI	Advanced Land Imager
ALT	Altimeter
AMSR	Advanced Microwave Scanning Radiometer
AMSR-E	Advanced Microwave Scanning Radiometer (EOS version)
AMSU	Advanced Microwave Sounding Unit
AO	Announcement of Opportunity
APL	Applied Physics Laboratory
ASTER	Advanced Spaceborne Thermal Emission and Reflection Radiometer
ATMS	Advanced Technology Microwave Sounder
AVHRR	Advanced Very High Resolution Radiometer
CERES	Clouds and the Earth's Radiation Energy System
CMIS	Conical Microwave Imager/Sounder
CPU	Central Processing Unit
CrIS	Cross-track Infrared Sounder
CZCS	Coastal Zone Color Scanner
DARPA	Defense Advanced Research Projects Agency
DCS	Data Collection System
DFA	Dual-Frequency Radar Altimeter
DMSP	Defense Meteorological Satellite Program
DOD	U.S. Department of Defense
EDR	Environmental Data Record
ENSO	El Niño-Southern Oscillation
EO-1	Earth Orbiter-1
EOS	Earth Observing System
EOSDIS	Earth Observing System Data and Information System
EOSAT	Earth Observation Satellite Company
ERBS	Earth Radiation Budget Sensor/Satellite

ESA	European Space Agency
ESE	Earth Science Enterprise
ETM+	Enhanced Thematic Mapper Plus
EUMETSAT	European Organisation for the Exploitation of Meteorological Satellites
FOO	Flight of Opportunity
FUSE	Far Ultraviolet Spectroscopic Explorer
GEM	Graphite Epoxy Motor
GLAS	Geoscience Laser Altimeter System
GLI	Global Imager
GOES	Geostationary Operational Environmental Satellite
GPS	Global Positioning System
GPSOS	Global Positioning System Occultation Sensor
GSFC	Goddard Space Flight Center
GTO	Geo-Transfer Orbit
HAPS	Hydrazine Auxiliary Propulsion System
HIRDLS	High Resolution Dynamics Limb Sounder
HSB	Humidity Sounder for Brazil
ICBM	Intercontinental Ballistic Missile
ICESat	Ice, Cloud, and Land Elevation Satellite
IDIQ	Indefinite Delivery, Indefinite Quantity
IELV	Intermediate-size Expendable Launch Vehicle
ILAS	Improved Limb Atmospheric Spectrometer
IPO	Integrated Program Office
JGOFS	Joint Global Ocean Flux Study
JMR	Jason Microwave Imager
JPL	Jet Propulsion Laboratory
Landsat	Land Remote Sensing Satellite
LEO	Low Earth Orbit
LIS	Lightning Imaging Sensor
LMLV-3	Lockheed Martin Launch Vehicle-3
Medlite	Medium-Light Expendable Launch Vehicle
MELV	Medium Expendable Launch Vehicle
MEPS	Medium-Energy Particle Spectrometer
METOP	Meteorological Operational Polar Orbiter
MIDEX	Mid-size Explorer
MISR	Multi-Angle Imaging Spectroradiometer
MLS	Microwave Limb Sounder
MLELV	Medium-Light Expendable Launch Vehicle (MedLite)
MODIS	Moderate-Resolution Imaging Spectrometer
MOPITT	Measurements of Pollution in the Troposphere
MTF	Modulation Transfer Function
MTPE	Mission to Planet Earth
NASA	National Aeronautics and Space Administration
NMP	New Millennium Program
NOAA	National Oceanic and Atmospheric Administration
NPOESS	National Polar-orbiting Operational Environmental Satellite System
NSCAT	NASA Scatterometer
NWS	National Weather Service
ODUS	Ozone Dynamics Ultraviolet Spectrometer
OMPS	Ozone Mapping and Profiler Suite

OSC	Orbital Sciences Corporation
PI	Principal Investigator
POES	Polar-orbiting Operational Environmental Satellite
POLDER	Polarization and Directionality of the Earth's Reflectance
PR	Precipitation Radar
QuikSCAT	Quick Scatterometer
R&A	Research and Analysis
RPA-D	Retarding Potential Analyzer-Driftmeter (or drift plasma sensor)
RPU	Repeater Processing Unit
RSDO	Rapid Spacecraft Development Office
SAGE	Stratospheric Aerosol and Gas Experiment
SARSAT	Search and Rescue Satellite Aided Tracking
SBRC	Santa Barbara Research Center
SeaWiFS	Sea-Viewing Wide Field-of-View Sensor
SELV	Small Expendable Launch Vehicle
SES	Space Environmental Suite/Sensor
SMEX	Small Explorer
SOLSTICE/SAVE	Solar-Stellar Irradiance Comparison Experiment/Solar Atmospheric Variability Explorer
SPARCLE	Space Readiness Coherent Lidar Experiment
SPOT	Systeme Pour l'Observation de la Terre
SPU	Signal Processing Unit
SST	Sea Surface Temperature
SSTI	Small Spacecraft Technology Initiative
STEP	Space Test Experiment Program
SWIR	Short-Wave Infrared
TDI	Time-Delay Integration
TES	Tropospheric Emission Spectrometer
TIR	Thermal Infrared
TMI	TRMM Microwave Imager
TOMS	Total Ozone Mapping Spectrometer
TOMS-EP	Total Ozone Mapping Spectrometer Earth Probe
TRMM	Tropical Rainfall Measuring Mission
TSI	Total Solar Irradiance
TSIS/M	Total Solar Irradiance Sensor/Monitor
VIIRS	Visible/Infrared Imaging/Radiometer Suite
VIRS	Visible Infrared Scanner
VNIR	Visible and Near Infrared